Computer Vision and Image Analysis for Industry 4.0

Computer vision and image analysis are indispensable components of every automated environment. Modern machine vision and image analysis techniques play key roles in automation and quality assurance. Working environments can be improved significantly if we can integrate computer vision and image analysis techniques. The more advancement of innovation and research in computer vision and image processing, the larger the increase in efficiency of machines as well as humans. *Computer Vision and Image Analysis for Industry 4.0* focuses on the roles of computer vision and image analysis for 4.0 IR-related technologies. This book proposes a variety of techniques for disease detection and prediction, text recognition and signature verification, image captioning, flood level assessment, crops classifications and fabrication of smart eye-controlled wheelchairs.

Computer Vision and Image Analysis for Industry 4.0

Edited by
Nazmul Siddique
Mohammad Shamsul Arefin
Md Atiqur Rahman Ahad
M. Ali Akber Dewan

CRC Press
Taylor & Francis Group
Boca Raton London New York

CRC Press is an imprint of the
Taylor & Francis Group, an **informa** business

A CHAPMAN & HALL BOOK

First edition published 2023
by CRC Press
6000 Broken Sound Parkway NW, Suite 300, Boca Raton, FL 33487-2742

and by CRC Press
4 Park Square, Milton Park, Abingdon, Oxon, OX14 4RN

CRC Press is an imprint of Taylor & Francis Group, LLC

Library of Congress Cataloging-in-Publication Data

Names: Siddique, N. H., editor. | Arefin, Mohammad Shamsul, editor. | Ahad, Md. Atiqur Rahman, editor. | Dewan, M. Ali Akber, editor.
Title: Computer vision and image analysis for industry 4.0 / edited by Nazmul Siddique, Mohammad Shamsul Arefin, Md Atiqur Rahman Ahad, and M. Ali Akber Dewan.
Description: First edition. | Boca Raton : CRC Press, 2023. | Includes bibliographical references and index.
Identifiers: LCCN 2022029347 (print) | LCCN 2022029348 (ebook) | ISBN 9781032164168 (hardback) | ISBN 9781032187624 (paperback) | ISBN 9781003256106 (ebook)
Subjects: LCSH: Computer vision. | Image analysis. | Image processing--Digital techniques. | Industry 4.0.
Classification: LCC TA1634 .C6488 2023 (print) | LCC TA1634 (ebook) | DDC 006.3/7--dc23/eng/20221024
LC record available at https://lccn.loc.gov/2022029347
LC ebook record available at https://lccn.loc.gov/2022029348

ISBN: 978-1-032-16416-8 (hbk)
ISBN: 978-1-032-18762-4 (pbk)
ISBN: 978-1-003-25610-6 (ebk)

DOI: 10.1201/9781003256106

Typeset in Latin Modern font
by KnowledgeWorks Global Ltd.

Publisher's note: This book has been prepared from camera-ready copy provided by the authors.

This book is dedicated to all authors who have contributed to this book.

Contents

Preface

We are in the era of the Fourth Industrial Revolution (4.0 IR), which is a new chapter in human development enabled by extraordinary technology advances and making a fundamental change in the way we live, work and relate to one another. It is an opportunity to help everyone, including leaders, policy-makers, and people from all income groups and nations, to harness converging technologies in order to create an inclusive, human-centered future. We need to prepare our students and professionals with 4.0 IR-related technologies.

Computer vision and image analysis plays an essential role in 4.0 IR as machine vision and image analysis are indispensable components of every automated environment. Modern machine vision and image analysis techniques play key roles in automation and quality assurance. Working environments can be improved significantly if we can integrate computer vision and image analysis techniques. More advancement in innovation and research in computer vision and image processing can increase the efficiency of machines as well as people. Considering these facts, in this book, we accommodate fifteen interesting chapters focusing on the roles of computer vision and image analysis for 4.0 IR-related technologies.

This book proposes a variety of techniques for disease detection and prediction, text recognition and signature verification, image captioning, flood level assessment, crop classifications and fabrication of smart eye controlled wheelchairs.

The book can be used as a reference for advanced undergraduate or graduate students and we believe that this book will play a vital role in enhancing the knowledge of young researchers, students and professionals in the domain of computer vision and image analysis. We are highly thankful to all the contributors of this book.

Best regards,

Nazmul Siddique
Mohammad Shamsul Arefin
Md Atiqur Rahman Ahad
M. Ali Akber Dewan

Contributors

Md. Ataur Rahman
Premier University, Dept. of CSE,
Chittagong, Bangladesh.

Nazifa Tabassum
Premier University, Dept. of CSE,
Chittagong, Bangladesh.

Mitu Paul
Premier University, Dept. of CSE,
Chittagong, Bangladesh.

Riya Pal
Premier University, Dept. of CSE,
Chittagong, Bangladesh.

Mohammad Khairul Islam
University of Chittagong, Dept. of CSE,
Chittagong, Bangladesh.

Nawmee Razia Rahman
Department of Computer Science and
    Engineering, RUET,
Rajshahi, Bangladesh

Md. Nazrul Islam Mondal
Department of Computer Science and
    Engineering, RUET,
Rajshahi, Bangladesh

S M Shahriar Sharif Rahat
Dept. of CSE, Begum Rokeya University,
Rangpur, Bangladesh

Manjara Hasin Al Pitom
Dept. of CSE, Begum Rokeya University,
Rangpur, Bangladesh

Mridula Mahzabun
Dept. of CSE, Begum Rokeya University,
Rangpur, Bangladesh

Md. Shamsuzzaman
Dept. of CSE, Begum Rokeya University,
Rangpur, Bangladesh

Mohiuddin Ahmed
Rajshahi University of Engineering &
    Technology,
Rajshahi, Bangladesh

Abu Sayeed
Rajshahi University of Engineering &
    Technology, Rajshahi,
Bangladesh

Azmain Yakin Srizon
Rajshahi University of Engineering &
    Technology,
Rajshahi, Bangladesh

Md Rakibul Haque
Rajshahi University of Engineering &
    Technology,
Rajshahi, Bangladesh

Md. Mehedi Hasan
Rajshahi University of Engineering &
    Technology,
Rajshahi, Bangladesh

Md Rakibul Haque
Rajshahi University of Engineering &
    Technology,
Rajshahi, Bangladesh

Azmain Yakin Srizon
Rajshahi University of Engineering &
    Technology,
Rajshahi, Bangladesh

Mohiuddin Ahmed
Rajshahi University of Engineering &
Technology,
Rajshahi, Bangladesh

Md. Shafayat Jamil
Computer Science and Engineering
Discipline, Khulna University,
Khulna, Bangladesh

Sirdarta Prashad Banik
Computer Science and Engineering
Discipline, Khulna University,
Khulna, Bangladesh

G. M. Atiqur Rahaman
Computer Science and Engineering
Discipline, Khulna University,
Khulna, Bangladesh

Sajib Saha
Australian e-Health Research Centre,
CSIRO,
Floreat, Australia

Thasin Abedin
Islamic University of Technology (IUT),
Gazipur, Bangladesh

Khondokar S. S. Prottoy
Islamic University of Technology (IUT),
Gazipur, Bangladesh

Ayana Moshruba
Islamic University of Technology (IUT),
Gazipur, Bangladesh

Safayat Bin Hakim
Islamic University of Technology (IUT),
Gazipur, Bangladesh

Ahmad Sabbir Chowdhury
Computer Science and Engineering,
Chittagong Independent University,
Chittagong, Bangladesh

Aseef Iqbal
Computer Science and Engineering,
Chittagong Independent University,
Chittagong, Bangladesh

Shamim Iben Shahid
National University of Science and
Technology MISiS,
Moscow, Russia

Mohammed Abdul Kader
Dept. of Electrical and Electronic
Engineering, International Islamic
University Chittagong,
Chattogram, Bangladesh

Muhammad Ahsan Ullah
Dept. of Electrical and Electronic
Engineering, Chittagong University of
Engineering and Technology,
Chattogram, Bangladesh

Md Saiful Islam
Dept. of Electronics and Telecommunication
Engineering, Chittagong University of
Engineering and Technology,
Chattogram, Bangladesh

SK. Shalauddin Kabir
Department of Computer Science and
Engineering, Jashore University of
Science and Technology,
Jashore, Bangladesh

Mohammad Farhad Bulbul
Department of Computer Science and
Engineering, Pohang University of
Science and Technology (POSTECH),
Pohang, Republic of Korea;
Department of Mathematics, Jashore
University of Science and Technology,
Jashore, Bangladesh

Fee Faysal Ahmed
Department of Mathematics, Jashore
University of Science and Technology,
Jashore, Bangladesh

Syed Galib
Department of Computer Science and
Engineering, Jashore University of
Science and Technology,
Jashore, Bangladesh

Hazrat Ali
College of Science and Engineering, Hamad
Bin Khalifa University,
Qatar Foundation, Doha, Qatar

Toshiba Kamruzzaman
Dept. of ECE, Rajshahi University of
Engineering and Technology,
Rajshahi, Bangladesh

Abdul Matin
Dept. of ECE, Rajshahi University of
Engineering and Technology,
Rajshahi, Bangladesh

Tasfia Seuti
Dept. of CSE, Rajshahi University of
Engineering and Technology,
Rajshahi, Bangladesh

Md. Rakibul Islam
Dept. of ECE, Rajshahi University of
Engineering and Technology,
Rajshahi, Bangladesh

Mohammad Golam Mortuza
Dept. of Electronics and
Telecommunication Engineering, CUET,
Chittagong, Bangladesh

Md. Humayun Kabir
Dept. of Electronics and
Telecommunication Engineering, CUET,
Chittagong, Bangladesh

Uipil Chong
University of Ulsan,
Ulsan, South Korea

Promiti Chakraborty
Department of Computer Science &
Engineering (CSE), CUET,
Chattogram, Bangladesh

Sabiha Anan
Department of Computer Science &
Engineering (CSE), CUET,
Chattogram, Bangladesh

Kaushik Deb
Department of Computer Science &
Engineering (CSE), CUET,
Chattogram, Bangladesh

Md. Anisur Rahman
Chittagong University of Engineering and
Technology,
Chattogram, Bangladesh

Md. Abdur Rahman
Chittagong University of Engineering and
Technology,
Chattogram, Bangladesh

Md. Imteaz Ahmed
Chittagong University of Engineering and
Technology,
Chattogram, Bangladesh

Md. Iftekher Hossain
Chittagong University of Engineering and
Technology,
Chattogram, Bangladesh

Editors

Nazmul Siddique is with the School of Computing, Engineering and Intelligent Systems, Ulster University. He obtained a Dipl.-Ing. degree in Cybernetics from the Dresden University of Technology, Germany, an MSc in Computer Science from Bangladesh University of Engineering and Technology, and a PhD in Intelligent Control from the Department of Automatic Control and Systems Engineering, University of Sheffield, England. His research interests include: cybernetics, computational intelligence, nature-inspired computing, stochastic systems and vehicular communication. He has published over 195 research papers including five books published by John Wiley, Springer and Taylor & Francis. He guest-edited eight special issues of reputed journals on Cybernetic Intelligence, Computational Intelligence, Neural Networks and Robotics. He is on the editorial board of seven international journals including *Nature Scientific Reports*. He is a Fellow of the Higher Education Academy, and a senior member of IEEE. He has been involved in organising many national and international conferences and co-edited seven conference proceedings.

Mohammad Shamsul Arefin is in leave from Chittagong University of Engineering and Technology (CUET) and currently affiliated with the Department of Computer Science and Engineering (CSE), Daffodil International University, Bangladesh. Earlier he was the Head of the Department of CSE, CUET. Prof. Arefin received his Doctor of Engineering Degree in Information Engineering from Hiroshima University, Japan, with support of the scholarship of MEXT, Japan. As a part of his doctoral research, Dr. Arefin was with IBM Yamato Software Laboratory, Japan. His research includes privacy preserving data publishing and mining, distributed and cloud computing, big data management, multilingual data management, semantic web, object oriented system development and IT for agriculture and environment. Dr. Arefin has more than 120 refereed publications in international journals, book series and conference proceedings. He is a senior member of IEEE, a Member of ACM, and a Fellow of IEB and BCS. Dr. Arefin is the Organizing Chair of BIM 2021; TPC Chair, ECCE 2017; Organizing Co-Chair, ECCE 2019; and Organizing Chair, BDML 2020. Dr. Arefin has visited Japan, Indonesia, Malaysia, Bhutan, Singapore, South Korea, Egypt, India, Saudi Arabia and China for different professional and social activities.

Md Atiqur Rahman Ahad, SMIEEE, SMOPTICA - is an Assoc. Prof., University of East London; Prof. (*former*), University of Dhaka (DU); and Specially Appointed Assoc. Prof. (*former*), Osaka University. He studied at the University of Dhaka (BSc(Honors) and Masters), University of New South Wales (Masters), and Kyushu Institute of Technology (PhD). He published 10+ books, 200+ journal/proceedings/chapters, delivered 140+ keynote/invited talks, and achieved 40+ Awards/Recognition. He is an *Editorial Board Member* of Scientific Reports, Nature; *Assoc. Editor* of Frontiers in Computer Science; *Editor* of Int. Journal of Affective Engineering; *Editor-in-Chief*: Int. Journal of Computer Vision and Signal Processing (IJCVSP); *Guest-Editor* of Pattern Recognition Letters, JMUI, JHE, IJICIC. He is a member of ACM and IAPR. More: `http://ahadVisionLab.com`.

M. Ali Akber Dewan, Member, IEEE received a B.Sc. degree in computer science and engineering from Khulna University, Bangladesh, in 2003, and a Ph.D. degree in computer engineering from Kyung Hee University, South Korea, in 2009. From 2003 to 2008, he was also a lecturer with the Department of Computer Science and Engineering, Chittagong University of Engineering and Technology, Bangladesh, where he was an Assistant Professor, in 2009. From 2009 to 2012, he was a Postdoctoral Researcher with Concordia University, Montreal, QC, Canada. From 2012 to 2014, he was a Research Associate with the École de Technologie Supérieure, Montreal. He is currently an Associate Professor at the School of Computing and Information Systems, Athabasca University, Canada. He has published more than 50 articles in peer reviewed journals and conference proceedings. His research interests include artificial intelligence, affective computing, computer vision, data mining, information visualization, machine learning, biometric recognition, medical image analysis, and health informatics. He has served as an editorial board member, a Chair/Co-Chair and a TPC member in several prestigious journals and conferences. He received the Dean's Award and the Excellent Research Achievement Award for his excellent academic performance and research achievements during his Ph.D. studies in South Korea.

CHAPTER 1

BN-HTRd: A Benchmark Dataset for Document Level Offline Bangla Handwritten Text Recognition (HTR) and Line Segmentation

Md. Ataur Rahman

Premier University, Dept. of CSE, Chittagong, Bangladesh

Nazifa Tabassum

Premier University, Dept. of CSE, Chittagong, Bangladesh

Mitu Paul

Premier University, Dept. of CSE, Chittagong, Bangladesh

Riya Pal

Premier University, Dept. of CSE, Chittagong, Bangladesh

Mohammad Khairul Islam

University of Chittagong, Dept. of CSE, Chittagong, Bangladesh

CONTENTS

W E introduce a new *dataset*[1] for offline Handwritten Text Recognition (HTR) from images of Bangla scripts comprising words, lines, and document-level annotations. The BN-HTRd dataset is based on the BBC Bangla News corpus, meant to act as ground truth texts. These texts were subsequently used to generate the annotations that were filled out by people with their handwriting. Our dataset includes 786 images of handwritten pages produced by approximately 150 different writers. It can be adopted as a basis for various handwriting classification tasks such as end-to-end document recognition, word-spotting, word or line segmentation, and so on. We also propose a scheme to segment Bangla handwritten document images into corresponding lines in an unsupervised manner. Our line segmentation approach takes care of the variability involved in different writing styles, accurately segmenting complex handwritten text lines of curvilinear nature. Along with a bunch of pre-processing and morphological operations, both Hough line and circle transforms were employed to distinguish different linear components. In order to arrange those components into their corresponding lines, we followed an unsupervised clustering approach. The average success rate of our segmentation technique is 81.57% in terms of *FM* metrics (similar to *F* measure) with a mean Average Precision (*mAP*) of 0.547.

1.1 INTRODUCTION

Data is the new oil in this era of the digital revolution. In order to make decisions through automatic and semi-automatic systems that employ machine learning (ML) and artificial intelligence (AI), we need to convert the handwritten documents in government, and non-government organizations, such as those in banks or that involve legal decision making. Although Bangla is one of the most highly spoken languages, so far, not too much attention has been given to the task of end-to-end handwritten text recognition of Bangla script documents. Because of the lack of document-level (full page) handwritten datasets, we are unable to make use of the capabilities of modern ML algorithms in this domain.

This chapter introduces the most extensive dataset named BN-HTRd, for Bangla handwritten images to support the advancement of end-to-end recognition of documents and texts. Our dataset contains a total of 786 full-page images collected from

---

[1]BN-HTRd Dataset: https://data.mendeley.com/datasets/743k6dm543

150 different writers. With a staggering 1,08,147 instances of handwritten words, distributed over 13,867 lines and 23,115 unique words, this is currently the largest and most comprehensive dataset in this field (see Table 1.1 for complete statistics). We also provide the `lines` and ground truth annotations for both `full-text` and `words`, along with the segmented images and their positions. The contents of our dataset come from a diverse news category (see Table 1.2), and annotators of different ages, genders and backgrounds, having variability in writing styles.

Table 1.1   Statistics of the Dataset.

Types	Counts
Number of writers	150
Total number of images	786
Total number of lines	13,867
Total number of words	1,08,147
Total number of unique words	23,115
Total number of punctuation's	7,446
Total number of characters	5,74,203

Table 1.2   Contents of Dataset According to News Categories.

Content Type	Documents	# of Pages
Sports	41	202
Coronavirus & Effected	29	164
Corona Treatment & Vaccine	16	90
Election	17	88
Story of a Lifetime	09	67
History	06	34
Political	04	24
Mission of Space	04	16
Corruption	04	13
Economy	04	10
Others	17	78

Segmenting document images into their most fundamental parts, such as words and text lines, is regarded as the most challenging problem in the domain of handwritten document image recognition, where the scripts are curvilinear in nature. Thus, we also present an unsupervised `segmentation` methodology of hand-written documents into corresponding `lines` along with our dataset. The proposed approach's main novelties consist of extending and combining some of the earlier reported works in the field of text line segmentation.

1.2   RELATED WORK

The task of handwriting recognition has captivated researchers for nearly a half-century. Although such initial triumph began with simple handwritten digit recognition, the first-ever massive character-level recognition task was arranged in 1992 by the First Census of Optical Character Recognition System Conference [1]. Eventually, researchers started building sentence-level [2] as well as document-level [3] offline handwritten datasets for English. The IAM dataset was subsequently used to initiate one of the most popular handwriting recognition shared tasks – ICDAR [4].

We found only a handful of Bangla handwriting datasets, among which the majority are isolated character datasets. One such dataset is BanglaLekha-Isolated [5], which is comprised of a set of 10 numerals, 50 basic characters, and 24 carefully curated compound characters. Each of the 84 character samples accumulated 2000 individual images. The resulting dataset incorporates a total of 1,66,105 images of handwritten characters after discarding the scribbles. It also holds information regarding the age and gender of the subjects from whom the writing samples were obtained. Another multipurpose handwritten character dataset named Ekush [6] consists of 3,67,018 characters. This dataset was collected from different regions of Bangladesh with equal numbers of male and female writers and varying age groups. The dataset contains a collection of modifiers as well, which is missing from other similar character-level datasets. The ISI [7] and CMATERdb [8] datasets are two of the oldest character based handwritten datasets for the Bangla language.

The only dataset that resembles our own dataset in terms of word-level annotation is the BanglaWriting [9] dataset. It includes the handwriting of 260 people of diverse ages and personalities. The authors used an annotation tool to annotate the pages with bounding boxes containing the words' Unicode representation. This dataset comprises a total of 32,787 characters and 21,234 words, and has a vocabulary size of 5,470. Although all of the bounding boxes of word labels were manually produced, they did not provide the actual ground truth of the pages from which the writings were generated. On top of that, almost all of the pages were short in length and can be comprehended as more like a paragraph instead of a full page document.

Proper segmentation of text lines and words from document images containing handwriting is an essential task before any kind of recognition, such as layout analysis and authorship identification, etc. It is still deemed as a challenging task due to the (i) irregular spacing between words and (ii) variations of writing habits among different authors. Hidden Markov Models (HMM) [10] were extensively used in continuous speech recognition and single-word handwriting recognition before the popularity of neural based models. As the parameter estimation of HMM is more general, it leads to better recognition results initially, even with fewer pre-processing operations that mainly dealt with positioning and scaling [11].

Other initial approaches to text line segmentation were mainly based on connected component (CC) analysis [12, 13]. In such scenarios, the connected components' average width and height are first estimated using some form of ad hoc or statistical methods. Different lines were separated by Hough Line transformation, and some form of clustering scheme allowed them to distinguish between components that fall in distinct lines [14]. A local region-based text-line segmentation algorithm was proposed in [15]. The lines are simply segmented by using a horizontal projection-based approach. Text regions are detected locally considering the corresponding approximated skew angles, and also considering skews of blocks, the proposed method outperformed all other approaches during that time.

In the Bangla script, a word can be horizontally partitioned into three adjoining zones – the lower zone, middle zone, and upper zone. The author of this article [16] used these zones in order to distinguish among different words within a line. Unfortunately, they only detected the words, which doesn't mean much without their location in the corresponding lines. Our approach relates to that work in that the black pixels on the 'matra' (contiguous upper zone) are automatically identified as segment points.

More advanced work in this domain for segmenting the text-line/word images into sub-words using a graph modeling-based approach was done in [17]. They achieved sub-word-level segmentation while also considering the issue of displacement of the diacritics. The use of convolutional neural networks with a combination of LSTM and other deep learning frameworks to detect and recognize the lines or words in an image became popular after 2017 [18, 19, 20, 21]. These promising results encouraged us to do more research in this direction. However, these techniques require a lot of training data. Our dataset is mainly targeted to achieve this in our future endeavors.

1.3 DATA ANNOTATION

Annotation is a means of populating a corpus by examining something in the world and then recording the observed characteristics. The dataset is essentially organized into a particular model that helps to process the needed information. In this section, we will briefly explain the different stages of our dataset creation and annotation process.

1.3.1 Data Collection and the Source

As a first step, we have collected individual text documents from BBC Bangla News[2] as our ground truth data by automatically Crawling/Scraping the website. We mainly preferred this source for our dataset because the BBC Bangla News does not require any restrictions and has an open access policy[3] to their data for the general public. In most cases, we downloaded both the TEXT and PDF file for a particular news item. Secondly, those pdf files and texts have been renamed according to a sequence (1 to 150) and placed in a separate folder (Fig. 1.1) before distributing them to different writers.

1.3.2 Data Distribution

For data annotation, we have to distribute data among individuals. So, we provided the folders (having the text and pdf) to people of different ages and professions. To be specific, 85 of those data were given to undergraduate students and the rest (65) to writers of different backgrounds. We also provided them with a sample (an annotated

---

[2]https://www.bbc.com/bengali
[3]https://www.bbc.com/bengali/institutional-37289190

folder) and annotation instructions. In return, they wrote the contents of the file and gave us back the images or hard copies of the handwritten pages. Note that a single writer had to write multiple pages (up to 20), in most cases because of the length of the news. Fig. 1.1 below represents the arrangements of our dataset for a single folder, and Fig. 1.2 shows a handwritten sample page.

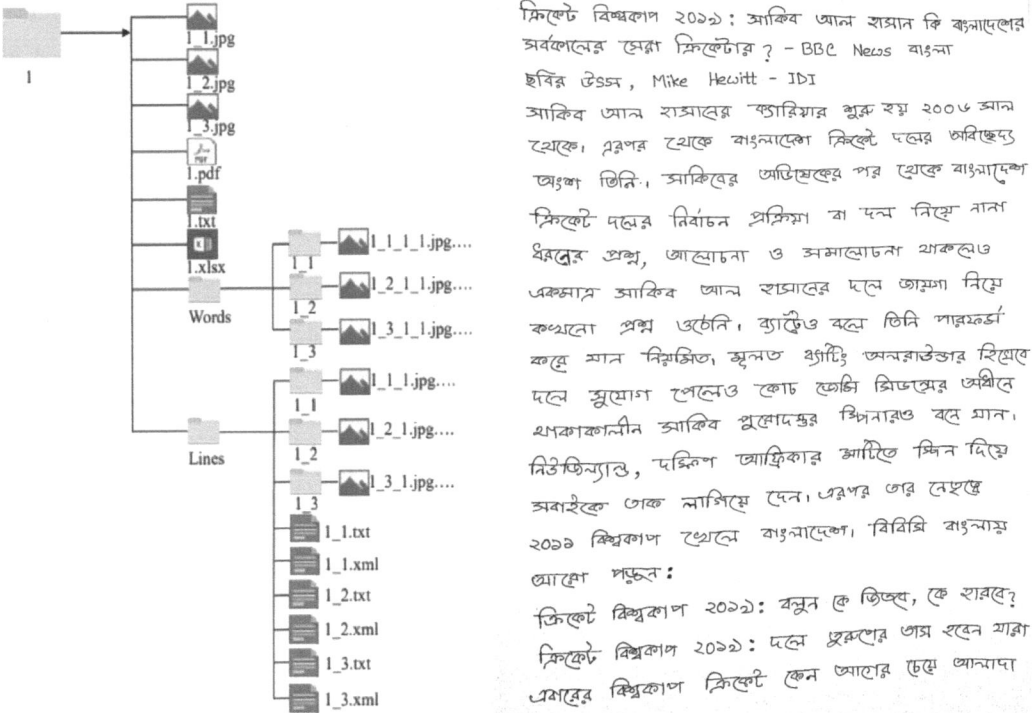

Figure 1.1 Dataset's Folder Structure. Figure 1.2 A Sample Image from the Dataset.

1.3.3 Annotation Guidelines

As the initial handwriting's were gathered from 150 native writers, each individual was instructed before writing the provided text to ensure that they wrote the text naturally. Thus, we provided the following guidelines to everyone:

1. Pages must be of A4 size and cannot be written on both sides.

2. If any line in the text file is missed while writing, it should be added at the end of the last page. Then the ground truth text was changed accordingly.

3. If the spelling of some word is wrong while writing, he/she should cut it with one pull, and the corrected one should be written next to it.

4. There is no need to end or start the line as it is in the text file. The following line can be started where the previous one ends.

5. While scanning the written text (page) with the CamScanner app, the camera's resolution should be good to ensure the image is not blurred.

1.3.4 Annotation Scheme and Agreement

We collected and digitized the handwritten pages into images and performed skew correction. After that, we segmented the pages into images of line, individual words (digits, punctuation) and created an Excel file for corresponding words. `Line` and `Word` folders are different for each image.

- Images are labeled by: `FolderNumber_PageNumber`.

- Lines are labeled by: `FolderNumber_PageNumber_LineNumber`.

- Words are labeled by: `FolderNumber_PageNumber_LineNumber_WordNumber`.

- Excel files are created including segmented words `Unicode` representation with their labeled ID.

The annotation of words was done in two folds. We instructed the 85 students to crop the words from the scanned pages that they wrote. The cropped images corresponded to each of the words and were placed into separate folders having their page and line number. We followed this specific pattern of naming the words/lines so that their filename corresponded to their position in the images (see above naming convention). Students also filled out an `EXCEL` file with the corresponding word identifier (file name in the aforementioned format) and their *Unicode Text* (Bangla word). The rest of the 65 authors' word annotations were done by us in a similar way.

For the line annotation, we used a tool named `LabelImg`[4] in order to get YOLO and PascalVOC formatted[5] annotations. Note that students did only the word segment part as an assignment. We did the line segmentation and all the other tasks related to the dataset curation. From those line annotations, we programmatically extracted the corresponding lines and arranged them in subfolders. We did this in order to apply deep learning frameworks such as YOLO/TensorFlow for line detection in a supervised way in our future research.

1.3.5 Data Correction

We made a video presentation for students to introduce the whole process of word annotation. After understanding the annotation process, students finally submitted their work, and we compared it word-by-word. Although their submission wasn't 100% accurate all the time, we corrected those faults manually. While examining, we tried to do it as much accurately as possible by following some rules such as:

1. The submitted images should match the given text file (line by line).

2. The cropped words (digits, punctuation) have been checked individually.

3. The ID and Word column of the Excel file have been manually verified.

4. The sequence of cropped words have been compared against the Excel file.

---

[4]LabelImg: `https://tzutalin.github.io/labelImg/`
[5]YOLO/PascalVOC Format: `https://cutt.ly/LvrTrCH`

1.4 LINE SEGMENTATION: METHODOLOGY

In this section, we describe our text line segmentation approach in detail. Fig. 1.3 illustrates the process of BN-HTRd dataset collection and preparation (left) and the overall system architecture of the line segmentation pipeline (right).

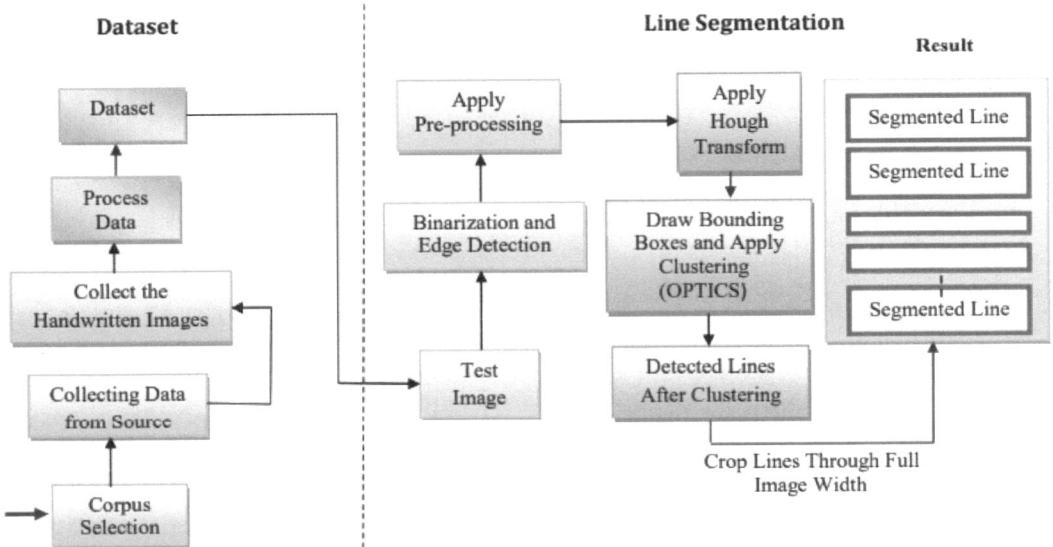

Figure 1.3 Overall System Architecture.

1.4.1 Thresholding and Edge Detection

Thresholding or image binarization is a non-linear process that transforms a grayscale (or colored) image to a binary image having only two levels (0 or 1) to represent each pixel considering the specified threshold value. In other words, if the pixel value is higher than the threshold, it is assigned one value (i.e., white); otherwise, it is assigned another value (i.e., black). We have incorporated a local OTSU technique to accomplish this. Fig. 1.4 shows the result of the binarization process over a segment of the original image.

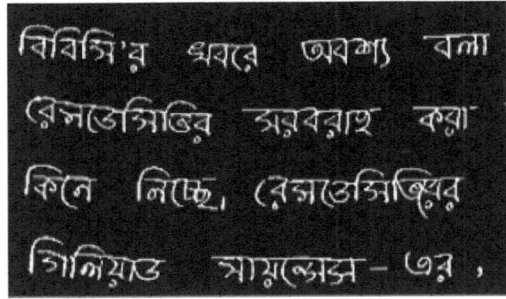

Figure 1.4 Thresholded Image.

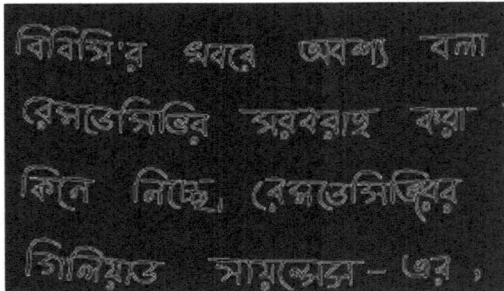

Figure 1.5 Canny Edge Detection.

Edge detection is a procedure that extracts or highlights useful regions in an image having different objects and subsequently reduces the number of non-useful pixels to be considered. We have used the Canny edge detection technique after binarization to highlight the edges of the handwritings (Fig. 1.5).

1.4.2 Morphological Operation and Noise Removal

In order to remove the small salt-and-pepper type noise, we used a morphological opening followed by dilation to separate the sure foreground (Fig. 1.6) noise from the background. To find the sure foreground objects, we used distance transformation and subtracted it from the background. The resultant image of Fig. 1.7 shows the situation after these preprocessing steps.

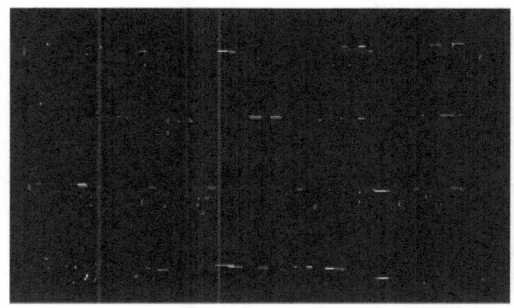
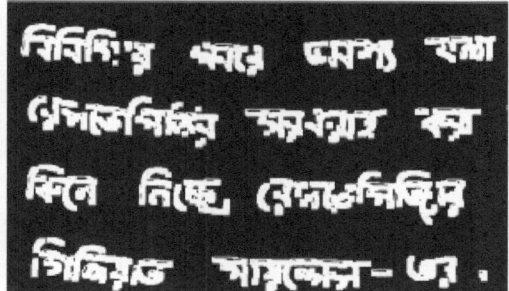

Figure 1.6 Sure Foreground (Noise).

Figure 1.7 After Removing Noise.

1.4.3 Hough Line Detection

We used the `Hough Line Transform` to distinguish the continuous horizontal lines ('matra') over the words and dilated them in order to thicken those lines so that each word acts as a connected component and separate words can be distinguished within a line. This will help us later on to draw the bounding box more accurately over the words. Fig. 1.8 shows the result.

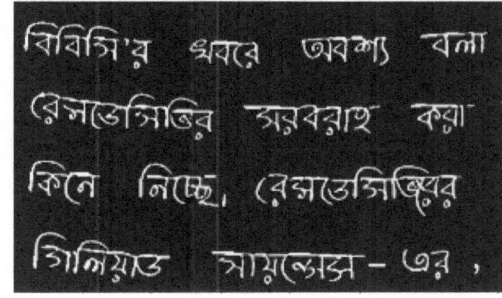
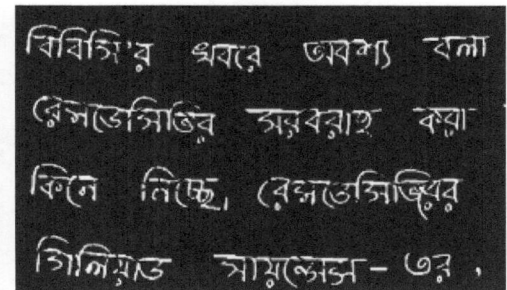

Figure 1.8 Hough Line Detection.

Figure 1.9 Hough Circle Removal.

1.4.4 Hough Circle Removal

Most of the time, two or more lines in a text document overlap due to the circle-like shape in Bangla scripts. We used the `Hough Circle Transform` to detect those circular objects and break them apart so that two consecutive horizontal lines don't form a connected component due to overlapping word segments. Fig. 1.9 illustrates the resultant partial image after the Hough circle removal operation.

1.4.5 Bounding Box

We used `Connected Component` (`CC`) analysis to draw bounding boxes over each connected region (Fig. 1.10). After that, we took boxes with a certain minimum area and determined the centre of the boxes (see the red dot inside the box) in order to use them in the next step for clustering.

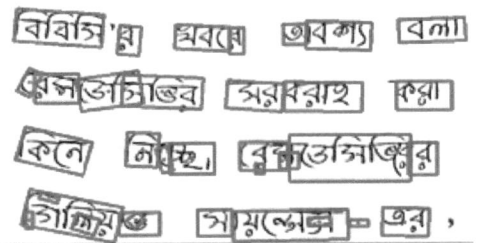

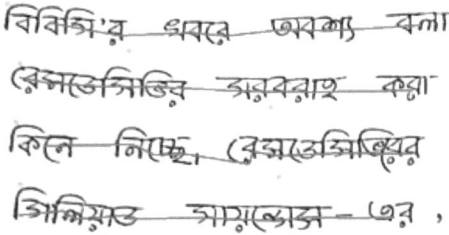

Figure 1.10 Bounding Box and Midpoints over the Connected Components.

Figure 1.11 Add the Midpoints and Marking Highest and Lowest Points of Line.

1.4.6 OPTICS Clustering

OPTICS Clustering stands for Ordering Points to Identify Cluster Structure. We used this algorithm over the Y co-ordinate (vertical axis) of the midpoints that we found in the previous step (section 1.4.5). It mostly gives us the points that fall in a line within a single cluster. Thus, after this operation, each cluster represents separate text lines in the image. After clustering, we added the midpoints and got the annotated line in Fig. 1.11. However, note that sometimes it fails to determine lines in cases when they are too close to each other or the midpoints of the bounding box are determined incorrectly, as shown in Fig. 1.12. One of the advantages of using OPTICS clustering is that, unlike K-Means clustering, we do not need to provide the number of clusters (k) beforehand. It serves our purpose as the number of lines per document image is not fixed and depends on the author's writing style and page size.

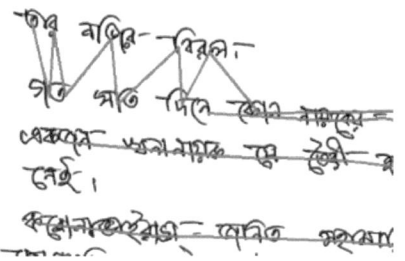

Figure 1.12 Failed to Cluster Properly.

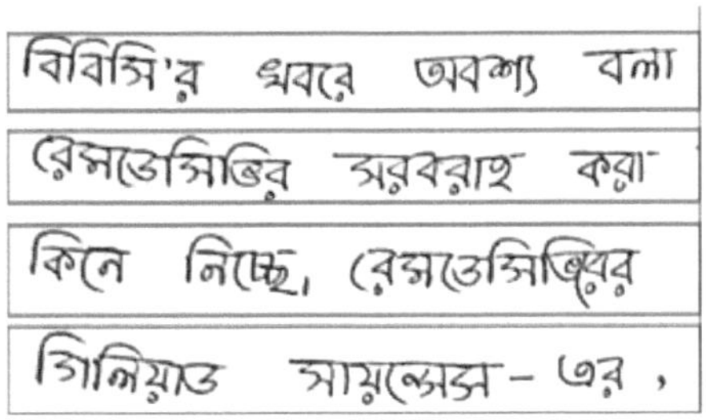

Figure 1.13 Bounding Box over the Lines

1.4.7 Line Extraction and Cropping

In order to visualize the lines, we used a rectangle around each of them (Fig. 1.13). We crop individual lines taking the full length of the bounding box. We considered the bounding boxes' top and bottom points (see green dots in Fig. 1.11) from the connected components (section 1.4.5) to determine the height of the cropped lines. Fig. 1.14 delineates two of the lines cropped through our method.

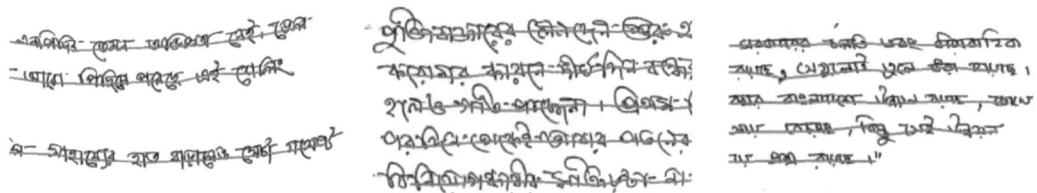

Figure 1.14 Two Cropped Lines from Fig 1.13.

1.5 RESULTS AND EVALUATION

In this section, we talk about our line segmentation results. Fig. 1.15 shows the output lines detected through our method for different handwritten images.

Figure 1.15 Output Lines for Different Types of Handwriting.

1.5.1 Evaluation Metrics

Two bounding boxes (lines) are considered as a one-to-one match if the total matching pixels is greater than or equal to the evaluator's approved threshold (T_a). Let N be the number of ground-truth elements, M be the count of detected components, and $o2o$ be the number of one-to-one matches between N and M. The Detection Rate (DR) and Recognition Accuracy (RA) are defined as follows:

$$DR = \frac{o2o}{N}, \quad RA = \frac{o2o}{M} \tag{1.1}$$

By combining the detection rate (DR) and recognition accuracy (RA), we can get the final performance metric FM (similar to F measure) using the equation below:

$$FM = \frac{2DR * RA}{DR + RA} \tag{1.2}$$

Average Precision (AP) in contrast, calculates the average values of Precision (P) for the corresponding Recall (R) values over 0 to 1 with an interval of 0.1:

$$AP = \frac{1}{11} \sum_{r \in 0,0.1,0.2,..1.0} P(r) \tag{1.3}$$

The mean Average Precision (mAP) is computed by using the mean AP across every class having a threshold (equivalent to T_a):

$$mAP = \frac{1}{n} \sum_{k=1}^{k=n} AP_k \qquad (1.4)$$

where AP_k is the average precision of class k, and n is the number of total classes.

1.5.2 Line Segmentation Results

We evaluated the performance of our algorithm for text line segmentation using equations 1.1–1.2 over a portion of our dataset[6] (150 images). The acceptance threshold used was $T_a = 80\%$. That is, if the bounding box of the ground truth line and our detected line have an 80% match in terms of the pixel area, we considered it accurate. Here, we only calculated the result for 150 unique images from 786 images (one image from every 150 folders). The total number of lines from 150 images are 2915, and by applying our method, we got 3437 lines, from which 2591 lines match with the ground truth having the aforementioned threshold. So, the value of N (ground truth) is 2915, the value of $o2o$ is 2591, and M is 3437. Now, using equations 1.1 and 1.2 we get:

$$DR = \frac{2591}{2915}, \quad RA = \frac{2591}{3437}, \quad FM = \mathbf{0.8157}$$

Apart from our own BN-HTRd dataset, we also tried the ICDAR2013 dataset[7] containing 50 Bangla test images for the Handwriting Segmentation Contest [12]. The results obtained from our unsupervised approach that we used to perform line segmentation for both datasets are presented in Table 1.3.

Table 1.3 Detailed Results on Two Datasets for Line Segmentation Based on FM (F Score).

Evaluation Metrics	Datasets	
	BN-HTRd	ICDAR2013
# of Images	150	50
N	2915	872
M	3437	943
o2o	2591	695
DR(%)	88.88%	79.7%
RA(%)	75.38%	73.7%
FM(%)	**81.57%**	76.58%

Table 1.4 Recall and Precision Values (11 point measurements).

No.	Recall	Precision
1.	1.0	1.0
2.	0.9	0.76
3.	0.8	0.73
4.	0.7	0.76
5.	0.6	0.71
6.	0.5	0.4
7.	0.4	0.42
8.	0.3	0.68
9.	0.2	0.3
10.	0.1	0.26
11.	0.0	0.0

---

[6]Line Segmentation Results: https://cutt.ly/cczzQ9i
[7]ICDAR2013 Dataset: https://cutt.ly/yvi8OrF

From the results, we can see that the `ICDAR2013` dataset contains fewer images having fewer lines as compared to our own dataset, thus M and o2o vary. Hence, there is a difference in terms of performance of our algorithm for these two datasets, which is nearly 5% (81.57% vs. 76.58%) in this case. Another reason is that the images in the `ICDAR2013` dataset are smaller in terms of resolution, which caused an intricacy as we have considered a standard resolution (width of at least 1000 pixels) while we run our system. To have a more accurate idea of the performance of our approach, we further calculated the recall and precision values for each of the 150 images that we tested from our `BN-HTRd` dataset. We then took the highest precision values (Table 1.4) for the recall values in the range 0.0−1.0 and having an interval of 1. Fig. 1.16 depicts this scenario in terms of a Recall vs. Precision graph. Here we only took 11 values since we only need these values for calculating AP and mAP.

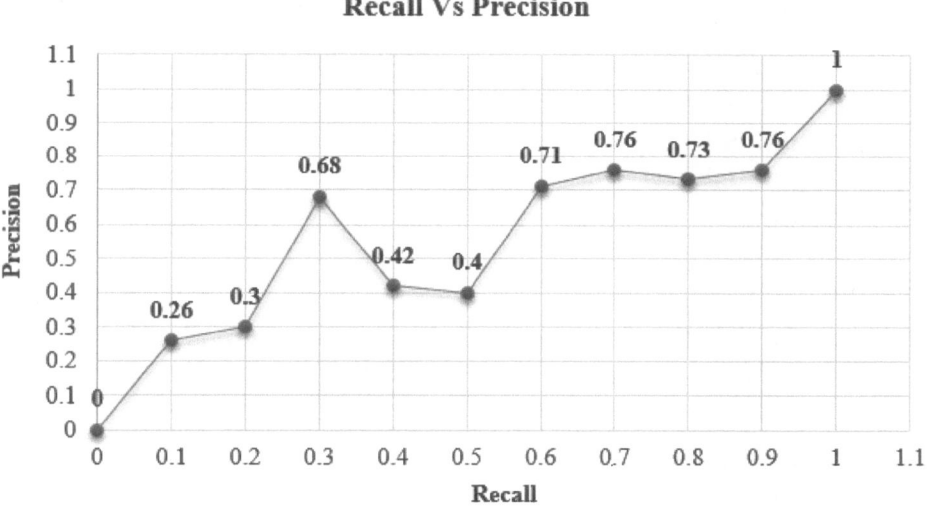

Figure 1.16 Recall v/s Precision Graph.

Using equation 1.3 and the values from Table 1.4, we get:

$$AP = \frac{1}{11}(1.0 + 0.76 + 0.73 + 0.76 + 0.71 + 0.4 + 0.42 + 0.68 + 0.3 + 0.26 + 0).$$

Thus the average precision for the `BN-HTRd` dataset is:

$$AP = \frac{1}{11}(6.02) = \mathbf{0.547}.$$

As we performed only the line segmentation, here we have only one class ($n = 1$ in equation 1.4). So, in this case, AP and mAP remain the same. Thus, our final `mAP` for line segmentation is **0.547**.

1.6   CONCLUSION AND FUTURE WORK

Our endeavour in this chapter was to lay the groundwork for future research on Bangla Handwritten Text Recognition (HTR). Keeping this in mind, we collected and developed the largest ever dataset in this domain, having both text line and word annotations as well as the ground truth texts for full-page handwritten document images. We also propose a framework for segmenting the lines from the input documents. Initially, the input images are resized and converted into binarized frames. Then the image noises and shaded effects are removed. After that, a connected component based segmentation method is applied to segment the components (mostly words) in the image. We employed OPTICS clustering on the bounding boxes from those segments in order to produce the final line segmentation. Our framework was able to achieve 81.57% in terms of FM score for line segmentation, having a mean average precision of 0.547 (mAP@0.8). In Bangla literature, many handwritten documents need to be converted into electronic versions. Line segmentation is an important part of achieving that goal. For that purpose, this work will help push the research on this direction one step forward. We aim to extend this work by incorporating word segmentation from the lines and recognizing individual words using deep learning models in the future. Our dataset is ready to deal with these objectives. We look forward to the research community around the world who will use this dataset to achieve the goal of end-to-end Bangla handwritten image recognition.

Bibliography

[1] Wilkinson, R. The first census optical character recognition system conference. (US Department of Commerce, National Institute of Standards, 1992)

[2] Marti, U. & Bunke, H. A full English sentence database for off-line handwriting recognition. *Proceedings of the Fifth International Conference on Document Analysis and Recognition. ICDAR'99 (Cat. No. PR00318).* pp. 705-708 (1999)

[3] Marti, U. & Bunke, H. The IAM-database: An English sentence database for offline handwriting recognition. *International Journal on Document Analysis and Recognition.* **5**, 39-46 (2002)

[4] Zimmermann, M. & Bunke, H. Automatic segmentation of the IAM off-line database for handwritten English text. *Object Recognition Supported by User Interaction for Service Robots.* **4** pp. 35-39 (2002)

[5] Biswas, M., Islam, R., Shom, G., Shopon, M., Mohammed, N., Momen, S. & Abedin, A. Banglalekha-isolated: A multi-purpose comprehensive dataset of handwritten Bangla isolated characters. *Data in Brief.* **12** pp. 103-107 (2017)

[6] Rabby, A., Haque, S., Islam, M., Abujar, S. & Hossain, S. Ekush: A multipurpose and multitype comprehensive database for online off-line Bangla handwritten characters. *International Conference on Recent Trends in Image Processing and Pattern Recognition.* pp. 149-158 (2018)

[7] Bhattacharya, U. & Chaudhuri, B. Handwritten numeral databases of Indian scripts and multistage recognition of mixed numerals. *IEEE Transactions on Pattern Analysis and Machine Intelligence.* **31**, 444-457 (2008)

[8] Sarkar, R., Das, N., Basu, S., Kundu, M., Nasipuri, M. & Basu, D. CMATERdb1: A database of unconstrained handwritten Bangla and Bangla–English mixed script document image. *International Journal on Document Analysis and Recognition (IJDAR).* **15**, 71-83 (2012)

[9] Mridha, M., Ohi, A., Ali, M., Emon, M. & Kabir, M. BanglaWriting: A multi-purpose offline Bangla handwriting dataset. *Data in Brief.* **34** pp. 106633 (2021)

[10] Rabiner, L. & Juang, B. A tutorial on hidden Markov models. *IEEE ASSP Magazine.* **3**, 4-16 (1986)

[11] Marti, U. & Bunke, H. Handwritten sentence recognition. *Proceedings 15th International Conference on Pattern Recognition. ICPR-2000.* **3** pp. 463-466 (2000)

[12] Stamatopoulos, N., Gatos, B., Louloudis, G., Pal, U. & Alaei, A. ICDAR 2013 handwriting segmentation contest. *2013 12th International Conference on Document Analysis and Recognition.* pp. 1402-1406 (2013)

[13] Ryu, J., Koo, H. & Cho, N. Language-independent text-line extraction algorithm for handwritten documents. *IEEE Signal Processing Letters.* **21**, 1115-1119 (2014)

[14] Ryu, J., Koo, H. & Cho, N. Word segmentation method for handwritten documents based on structured learning. *IEEE Signal Processing Letters.* **22**, 1161-1165 (2015)

[15] Ziaratban, M. & Bagheri, F. Extracting local reliable text regions to segment complex handwritten textlines. *2013 8th Iranian Conference on Machine Vision and Image Processing (MVIP).* pp. 70-74 (2013)

[16] Basu, S., Sarkar, R., Das, N., Kundu, M., Nasipuri, M. & Basu, D. A fuzzy technique for segmentation of handwritten Bangla word images. *2007 International Conference on Computing: Theory and Applications (ICCTA'07).* pp. 427-433 (2007)

[17] Ghaleb, H., Nagabhushan, P. & Pal, U. Segmentation of offline handwritten Arabic text. *2017 1st International Workshop on Arabic Script Analysis and Recognition (ASAR).* pp. 41-45 (2017)

[18] Renton, G., Chatelain, C., Adam, S., Kermorvant, C. & Paquet, T. Handwritten text line segmentation using fully convolutional network. *2017 14th IAPR International Conference on Document Analysis and Recognition (ICDAR).* **5** pp. 5-9 (2017)

[19] Bluche, T., Kermorvant, C., Ney, H., Bezerra, B., Zanchettin, C. & Toselli, A. How to design deep neural networks for handwriting recognition. *Handwriting: Recognition, Development and Analysis.* pp. 113-148 (2017)

[20] Bluche, T., Louradour, J. & Messina, R. Scan, attend and read: End-to-end handwritten paragraph recognition with mdlstm attention. *2017 14th IAPR International Conference on Document Analysis And Recognition (ICDAR).* **1** pp. 1050-1055 (2017)

[21] Bluche, T., Primet, M. & Gisselbrecht, T. Small-footprint open-vocabulary keyword spotting with quantized LSTM networks. *ArXiv Preprint ArXiv:2002.10851.* (2020)

CHAPTER 2

A New Approach Using a Convolutional Neural Network for Crop and Weed Classification

Nawmee Razia Rahman

Department of Computer Science and Engineering, Rajshahi University of Engineering & Technology, Rajshahi, Bangladesh

Md. Nazrul Islam Mondal

Department of Computer Science and Engineering, Rajshahi University of Engineering & Technology, Rajshahi, Bangladesh

CONTENTS

WEed control is a difficult issue that can affect crop productivity. Weeds are regarded as a serious issue because they reduce crop yields by increasing competition for nutrients, water, sunlight, and act as hosts for diseases and pests. It's critical to spot weeds early in their life cycle in order to prevent their grim effects on crop development. For this purpose, the aim of this chapter is to create a new crop and weed classification model. The Aarhus University Signal Processing group collaborated with the University of Southern Denmark to create a public dataset of 5,539 plant photographs of twelve types of crops and weeds. To classify these images effectively, previously several convolutional neural network (CNN) models have been developed with an overall classification accuracy of 90.15% and 94.38%. In this chapter, we propose a new and simple model using CNN considering the same dataset. The proposed model performs slightly better with 95.31% classification accuracy than that of other models developed earlier.

2.1 INTRODUCTION

Agriculture is considered to be the backbone of many economies around the world, especially in developing countries. Farmers are facing new threats as a result of population growth and global warming. On the other side, they must increase agricultural productivity. Therefore, we need to do agriculture more intelligently to reduce the losses. New technologies assist in solving these problems. This revolution led to the rise of a new term: Precision Agriculture or Precision Farming.

Precision Agriculture is the use of technological innovations to improve agricultural methods by enabling farmers to better understand all the data associated with their farms. It involves the use of a Global Positioning System (GPS), sensors, weather tracking, and other advancements. Farmers will gain powerful insights from all of these materials, which will aid them in making decisions. The ability to recognize weeds among native plant species is one of the most challenging aspects of farming. This task is typically carried out by field personnel. But, due to their strong similarities, it is difficult to detect and distinguish weeds from crops. Weeds directly compete with the crop for nutrients and water during the first six to eight weeks after seeding. As a result, cultivated crop yields suffer economic loss. Loss due to weeds varies depending on the type of weed, the type of crop, and the surrounding environmental conditions. It is essential to reliably detect weeds to minimize losses and, as a result, improve efficiency. There are approximately 30,000 species of plants known as weeds. Among these, 250 are classified in crop production as very troublesome, which causes major losses in yield.

Increasing weed competition decreases seed germination, and growth and development parameters such as plant height, dry matter accumulation, leaf area index, leaf thickness are reduced at the later stage of crop growth. Likewise, increasing competition from weeds dramatically decreases crop yields by around 45% [1]. Artificial intelligence advancements have significantly helped in the identification of weeds. Deep Learning is part of artificial intelligence. A neural network follows the working pattern of the brain. Any neural network that is more than two layers deep is known as a deep neural network, also called deep learning. Deep Learning networks learn higher-level functions from data on their own, automatically.

The convolutional Neural Network (CNN) is a form of neural network architecture that is very common [2]. They're mainly great at image processing [3], however, they're still good at a lot of other things such as speech recognition [4], natural language processing [5], in radiology [6] and more [7], [8]. Since most CNNs have a number of layers (at least dozens), they're called Deep Learning.

Previously, there have been several research studies in the field of agriculture using convolutional neural networks. For plant leaf identification [9] and disease identification [10] CNN is a commonly used approach. In this work, we used Convolutional Neural Networks to classify crop and weed images. The main contributions of this chapter are as follows:

- We propose a simple and new model using a CNN in order to enhance productivity for classifying crops and weeds effectively.

- As our model provides higher accuracy, this model can be implemented in an IoT based system that can assist farmers effectively.

2.2 CONVOLUTIONAL NEURAL NETWORK

A neural network is a set of interconnected artificial "neurons" that communicate with one another. The networks have numeric weights that are tuned during the training process so that when faced with an image or pattern to identify, a properly trained network can respond correctly. Many layers of feature-detecting "neurons" make up the network. Each layer contains a large number of neurons that relate to a variety of inputs from the previous layers. The layers are designed in such a way that if the first layer detects a series of primitive patterns in the input, the second layer detects variations of patterns, the third layer identifies patterns of those patterns, and so on. The training is conducted with a "labeled" dataset of inputs in a variety of representative input patterns that are labeled with the expected output response. The weights for intermediate and final function neurons are determined iteratively using general-purpose methods.

CNNs are comprised of three types of layers. These are convolutional layers, pooling layers, and fully connected layers. When these layers are stacked, a CNN architecture has been formed. The basic functionality of the CNN above can be broken down into four key areas.

a. Input Layer: As found in other forms of Artificial Neural Network, the input layer will hold the pixel values of the image.

b. Convolutional Layer: A convolutional layer includes a series of filters that learn about the parameters. The filters are lower in height and weight than those of the input volume. To compute an activation map made of neurons, each filter is convoluted with the input volume. In other words, at each spatial position, the filter is slid across the width and height of the input, and the dot products between the input and filter are computed. By stacking the activation maps of all filters along the depth axis, the output volume of the convolutional layer is obtained. As the width and height of each filter is built to be smaller than the input, only a small local area of the input volume is connected to each neuron in the activation map. The activation map is obtained by a convolution between the filter and the input, and all local positions are shared with the filter parameters. Weight sharing decreases the number of parameters for learning effectively.

c. Pooling Layer: To decrease the dimensions of the activation maps or feature maps, pooling layers are used. It thus decreases the number of learning parameters and the amount of computation performed by the network. There are several types of pooling layers. A pooling procedure that takes the maximum element of the filter-covered feature map region is defined as max-pooling. After the max-pooling layer, the result will be a feature map comprising the most prominent features from the preceding feature map. The average of the items present in the filter-covered feature map region is calculated using average pooling. As a result, while max-pooling returns the feature map's most prominent feature, average pooling returns the feature map's average.

d. The fully connected layer: A fully connected layer's goal is to take the convolution/pooling process results and use them to classify the image into a label. For example, if the picture is of a cat, the label "cat" should have a high likelihood of features reflecting items like whiskers or fur. Figure 2.1 is an example of a typical CNN architecture.

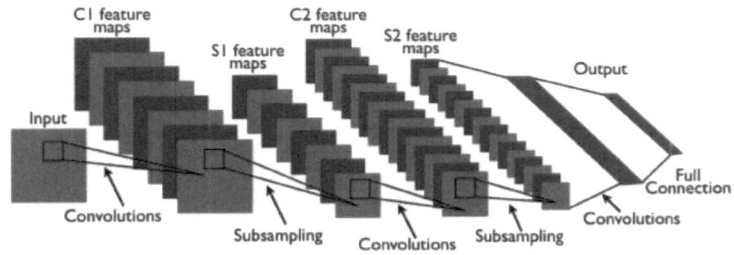

Figure 2.1   A typical CNN architecture [11].

2.3   THE PROPOSED MODEL

To perform the classification task in this work we constructed a simple model. It has only thirteen layers. Among the thirteen layers, there are seven convolutional layers, four pooling layers and two dense layers. The first convolutional layer has 32 filters and a 3x3 size kernel is used. The second convolutional layer uses similar properties as the first convolutional layer. The third layer is a pooling layer. The pooling operation is performed by maxpooling operation and a pool size of 2x2 is used. The fourth and fifth layers are two convolutional layers each having 64 filters with a kernel size of 3x3. The sixth layer is a maxpooling layer with a pool size of 2x2. The seventh and the eighth layers are two convolutional layers with 128 filters and a kernel size of 3x3. The ninth layer is a maxpool layer with pool size of 2x2. The tenth layer is a convolutional layer with 256 filters and kernel size 3x3. The eleventh layer is a maxpooling layer with pool size 2x2. The twelfth layer is a dense layer with 1024 nodes and the thirteenth layer is also a dense layer with 12 nodes. This last dense layer is the output layer. Each convolutional layer is followed by ReLU activation function and batch normalization. Each maxpool layer is followed by dropout of .25 to reduce overfitting. This architecture can be better understood from Figure 2.2.

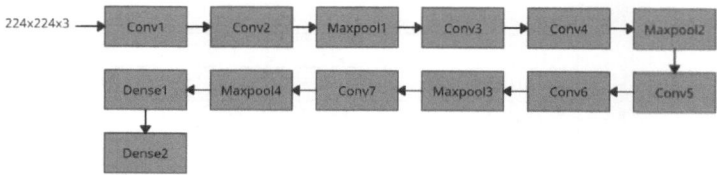

Figure 2.2 Architecture of our proposed model using CNN.

2.4 METHODOLOGY

In this section, dataset preparation and the whole learning procedure has been described.

2.4.1 Data Source

The used database was recorded at Aarhus University Flakkebjerg Research station in a collaboration between the University of Southern Denmark and Aarhus University and made public so that researchers can train their models on this dataset. It is available at [12].

2.4.2 Dataset Description

The Plant Seedlings Dataset contains images of approximately 960 unique plants belonging to 12 species at several growth stages. It comprises annotated RGB images.

Crops and weeds were observed for 20 days, with pictures taken every 2-3 days to capture different stages of development. There are 5,539 non-segmented images in this dataset. Black-grass, Charlock, Cleavers, Common Chickweed, Fat Hen, Loose silky-bent, Scentless-Mayweed, Shepherd's Purse, Small-flowered Cranesbill are the weeds and Common Wheat, Maize and Sugar Beet are the crops. Weeds make up the majority of the classes. This dataset was included in version two. The class distribution is presented in Table 2.1.

2.4.3 Work Procedure

To perform this classification task accurately and in less time, the whole task is divided into two parts. The first step is dataset preprocessing, and the second step is classifying the dataset with our proposed model. Deep learning architectures alone are robust enough to perform this classification task. But to increase accuracy, several image processing techniques are adopted. In Figure 2.3 the steps that are followed for the classification task are presented. The crop-weed dataset is the input to the system. Several image pre-processing steps are followed. Then these images are passed to the CNN model and the classification task is performed.

Table 2.1 Table of class distribution

Specie	Number of images
Black-grass	309
Charlock	452
Cleavers	335
Common Chickweed	713
Common Wheat	253
Fat Hen	538
Loose silky-bent	762
Maize	257
Scentless-Mayweed	607
Shepherd's Purse	274
Small-flowered Cranesbill	576
Sugar beet	463
Total	5,539

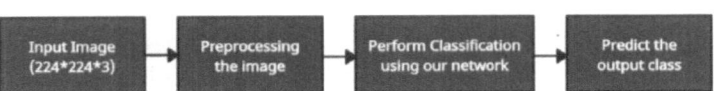

Figure 2.3 Steps followed in the classification task

2.4.4 Data Preprocessing

Data preprocessing is mainly done to increase accuracy and to reduce overfitting of the model. First, all the images are resized to 224x224 pixels so that all the networks take the same input. The images are resized in a range of [0-1] to reduce the difference between the pixel values. The dataset is imbalanced, so several augmentation techniques are used. The selected augmentation techniques are: rotation range of - 20° to 20°, zooming in a range of 0-10%, height-shift, and width-shift in a range of 0-10%, and lastly, horizontal flip.

2.4.5 Experimental Setup and Evaluation Metrics

There are 4436 data for training, 549 data for validation, and 554 data to evaluate the performance of the architecture after a 80/10/10 split. 150 epochs are used for training. For improved learning, the batch size is set to 20 with a learning rate of .0001, and Adam is used as an optimization function. Since data is multi-class, categorical cross-entropy is used as a loss function. The calculated performance matrix is:

$$Accuracy = (TP + TN)/(TP + TN + FP + FN) \qquad (2.1)$$

$$Precision = TP/(TP + FP) \qquad (2.2)$$

$$Recall = TP/(TP + FN) \qquad (2.3)$$

$$F1Score = 2x((Precision * Recall)/(Precision + Recall)) \qquad (2.4)$$

2.5 RESULT AND DISCUSSION

Our proposed model is tested on the dataset described in the previous section. The training set is used for training, the validation set is used to tune hyper parameters to prevent over fitting, and the test set is used to calculate the final performance of the model. The result we have achieved is 95.31% overall accuracy, 95.55% precision, 95.47% recall and 95.48% f1-score. We present the classification report of each class in Table 2.2.

Table 2.2 Table of classification report

Class	Precision	Recall	F1-score
Black-grass	.87	.87	.87
Charlock	.98	.95	.96
Cleavers	1	.94	.97
Common Chickweed	.95	.98	.97
Common Wheat	.91	.95	.93
Fat Hen	.95	1	.98
Loose silky-bent	.93	.90	.92
Maize	.96	1	.98
Scentless-Mayweed	.97	.97	.97
Shepherd's Purse	.95	.82	.88
Small-flowered Cranesbill	.98	.98	.98
Sugar beet	.96	1	.98

We also calculated the confusion matrix of the architecture. It is presented in Figure 2.4. From the figure of the confusion matrix, we can see how many classes are actually predicted by our model and the number of misclassifications. From the matrix, the following facts are observed:

- **Black-grass**: 33 images were predicted correctly among 38 images.

- **Charlock**: 53 images were predicted correctly among 56 images.

- **Cleavers**: 30 images were predicted correctly among 32 images.

- **Common Chickweed**: 62 images were predicted correctly among 63 images.

- **Common wheat**: 21 images were predicted correctly among 22 images.

- **Fat Hen**: 59 images were predicted correctly among 59 images.

- **Loose Silky-bent**: 66 images were predicted correctly among 73 images.

- **Maise**: 23 images were predicted correctly among 23 images.

- **Scentless Mayweed**: 64 images were predicted correctly among 66 images.

- **Shepherd's Purse**: 18 images were predicted correctly among 22 images.

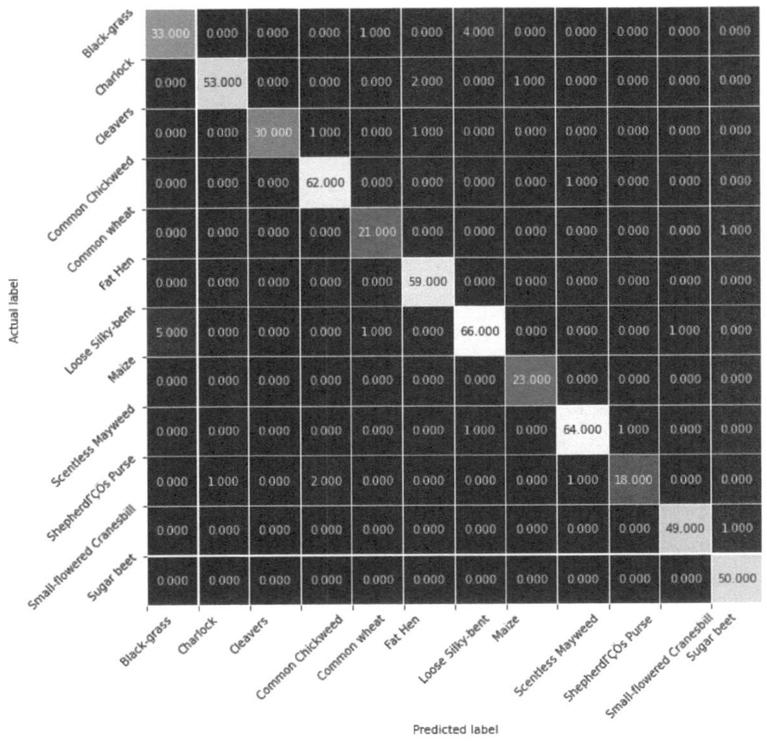

Figure 2.4 Confusion matrix of our proposed architecture.

- **Small flowered Cranesbill**: 49 images were predicted correctly among 50 images.

- **Sugar beet**: 50 images were predicted correctly among 50 images.

For better understanding of the performance of the model, we also calculated the accuracy and loss curves. The accuracy curve is presented in Figure 2.5. The loss curve is presented in Figure 2.6. From the loss curve it is seen that as the epoch increases, the loss decreases and training and validation loss almost align with each other, which is a sign of better fit.

The result of our architecture has been compared with other proposed architectures and the results is presented in Table 2.3. From this table we can see that our architecture performs better compared to other recent works.

Table 2.3 Table of result comparison

Source	Year	No. of Images	Accuracy
Alimboyong and Hernandez [13]	2019	4,234	90.15%
Elnemr [14]	2019	5,539	94.38%
Our proposed architecture	2021	5,539	95.31%

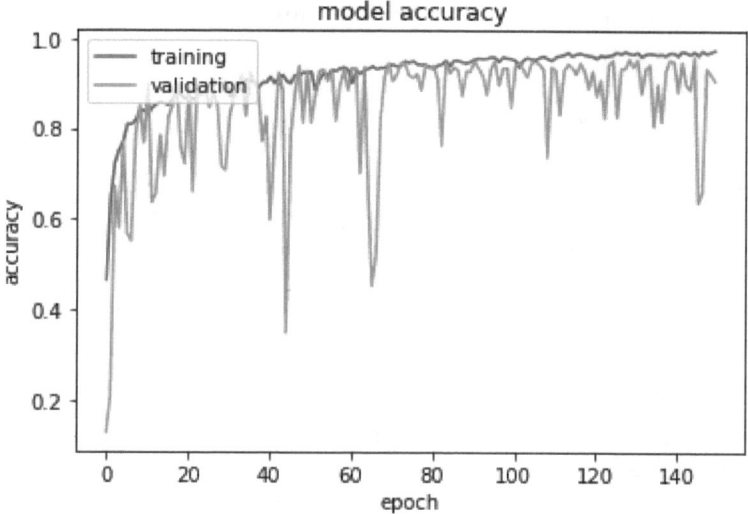

Figure 2.5 The accuracy curve of our proposed architecture.

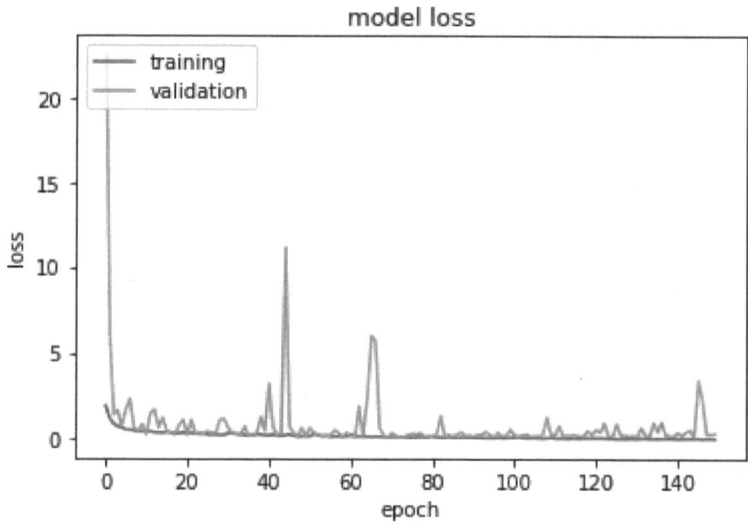

Figure 2.6 The loss curve of our proposed architecture.

2.6 CONCLUSION

This chapter presents a simple and new model CNN for classifying crops and weeds effectively for the farmer's benefit. The aim of this research is to create a deep neural network that can distinguish between plant seedling images and weeds. The network is designed from scratch. Then we look at how well a fully trained network can identify seedling images of twelve different species. With an accuracy rate of 95.31%, our proposed model performs exceptionally well in comparison to other architectures as mentioned. In the future, we have a plan to implement our trained proposed model in an IoT based system that can assist farmers.

Bibliography

[1] Reddy and Chandramohan. A study on crop weed competition in field crops. *Journal of Pharmacognosy and Phytochemistry*, 7(4):3235–3240, 2018.

[2] Saad Albawi, Tareq Abed Mohammed, and Saad Al-Zawi. Understanding of a convolutional neural network. International Conference on Engineering and Technology (ICET), pages 1–6. IEEE, 2017.

[3] Boukaye Boubacar Traore, Bernard Kamsu-Foguem, and Fana Tangara. Deep convolution neural network for image recognition. Ecological Informatics, 48:257–268, 2018.

[4] Zhengwei Huang, Ming Dong, Qirong Mao, and Yongzhao Zhan. Speech emotion recognition using CNN. In Proceedings of the 22nd ACM International conference on Multimedia, pages 801–804, 2014.

[5] Tom Young, Devamanyu Hazarika, Soujanya Poria, and Erik Cambria. Recent trends in deep learning based natural language processing. IEEE Computational Intelligence Magazine, 13(3):55–75, 2018.

[6] Rikiya Yamashita, Mizuho Nishio, Richard Kinh Gian Do, and Kaori Togashi.Convolutional neural networks: An overview and application in radiology. Insights into Imaging, 9(4):611–629, 2018.

[7] Yan Zhu, Qiu-Cheng Wang, Mei-Dong Xu, Zhen Zhang, Jing Cheng, Yun-ShiZhong, Yi-Qun Zhang, Wei-Feng Chen, Li-Qing Yao, Ping-Hong Zhou, et al. Application of convolutional neural network in the diagnosis of the invasion depth of gastric cancer based on conventional endoscopy. Gastrointestinal endoscopy, 89(4):806–815, 2019.

[8] Yao Zhang, Woong Je Sung, and Dimitri N Mavris. Application of convolutional neural network to predict airfoil lift coefficient. In 2018 AIAA/ASCE/AHS/ASC Structures, Structural Dynamics, and Materials Conference, page 1903, 2018.

[9] Christoph Wick and Frank Puppe. Leaf identification using a deep convolutional neural network. arXiv preprint arXiv:1712.00967, 2017.

[10] S Arivazhagan and S Vineth Ligi. Mango leaf diseases identification using convolutional neural network. International Journal of Pure and Applied Mathematics, 120(6):11067–11079, 2018.

[11] Yann LeCun, Koray Kavukcuoglu, and Clement Farabet. Convolutional nerworks and applications in vision. In Proceedings of 2010 IEEE International Symposium on Circuits and Systems, pages 253–256. IEEE, 2010.

[12] Plant seedlings dataset – Computer vision and bio systems signal processing groups. [https://vision.eng.au.dk/plant-seedlings-dataset/].

[13] Catherine R Alimboyong and Alexander A Hernandez. An improved deep neural network for classification of plant seedling images. In 2019 IEEE 15th International Colloquium on Signal Processing & Its Applications (CSPA), pages 217–222. IEEE, 2019.

[14] Heba A Elnemr. Convolutional neural network architecture for plant seedling classification. International Journal of Advanced Computer Science and Applications, 10(8):319–325, 2019.

CHAPTER 3

Lemon Fruit Detection and Instance Segmentation in an Orchard Environment Using Mask R-CNN and YOLOv5

S M Shahriar Sharif Rahat

Dept. of CSE, Begum Rokeya University, Rangpur, Bangladesh

Manjara Hasin Al Pitom

Dept. of CSE, Begum Rokeya University, Rangpur, Bangladesh

Mridula Mahzabun

Dept. of CSE, Begum Rokeya University, Rangpur, Bangladesh

Md. Shamsuzzaman

Dept. of CSE, Begum Rokeya University, Rangpur, Bangladesh

CONTENTS

This chapter presents the use of models that can identify, localize, and estimate the number of lemons from any commercial lemon orchard. Object detection, separation, and estimation were carried out by using Mask R-CNN and YOLOv5 machine learning models. These models were trained on images collected from different commercial lemon orchards around Bangladesh. According to the Mask R-CNN and YOLOv5 models, the FPR and FNR were 12.7% and 15.8%, and 7.2% and 9.4%, respectively. For the Mask R-CNN and YOLOv5 models, F1 scores were obtained as 85.6% and 91.5%, respectively. This estimation is comparable to other fruit detection models available in literature. This study can benefit commercial lemon producers in estimating the production of lemons from the viewpoint of agricultural requirements such as soil management, fertilizer use, and crop nutrition. The main challenges of this project were collecting images from the lemon orchards and annotating images to create the datasets. This study can help farmers and agricultural corporations accurately estimate the production of lemons.

3.1 INTRODUCTION

Lemon is a native South Asian evergreen tree from the citrus or rue family. The scientific name of lemon is *Citrus Limon*. This fruit is primarily used for its juice which contains a high amount of vitamin C of approximately 31 mg per lemon [1]. Lemons can provide 51% of the necessary daily vitamin C intake for humans [1]. In addition to its great health benefits, lemon can reduce the chance of heart and kidney disease, anemia, and digestive problems [1]. However, in Bangladesh, the average intake of vitamin C necessary for the human body is still below the recommended level. It is estimated that almost 93% of the population of Bangladesh are suffering from vitamin C deficiency [2].

In 1970, the annual lemon production in Bangladesh was estimated as 23,513 tons, which was increased to about 165,327 tons in 2019 [3]. Annually, the average production of lemons has increased by 5.06% with an export value in 2020 of US$328,000 [4]. Hence, lemon production in Bangladesh has become one of the promising sectors that is growing steadily. In addition, the demand for lemon suddenly increased due to the rise of the recent COVID-19 pandemic as doctors are suggesting to increase the daily vitamin C intake in order to stay healthy.

Recently, labor has become one of the most expensive aspects in the agricultural sector because of the increasing price for essential supplies such as electricity, irrigation, fertilizers, repellents, and pesticides. As a result, farm business owners and the horticultural industry are facing many challenges to meet the increasing cost of production. Despite these challenges, food production must continue to meet the rising demands of an ever-increasing global population. To solve these problems, the use of agricultural robots and machinery to harvest fruits and vegetables has now become necessary. In this regard, intelligent fruit detection techniques have evolved as a major research category in the robot harvesting field and successfully used to estimate different types of fruits such as passion fruit [5], apples [6] [7], cucumbers [8], mangoes [7] [9], and waxberries [10].

The aim of this study is to create new datasets and compare the experimental results of the Mask R-CNN and YOLOv5 models. We prepared two complete datasets of lemon images by collecting images from different orchards and annotated them separately for the two models. Finally, we compared the accuracy of these two models with the literature. Both the datasets consist of 700 images each (500 training images, 100 validation images, and 100 test images).

3.2 LITERATURE REVIEW

3.2.1 Texture, color, and shape based fruit detection

The algorithms of earlier fruit detection techniques are limited to texture and color of the fruits, and researchers mainly rely on the color contrast. When the color of fruits are different from the color of leaves, the location of a fruit is identified using threshold and pixel detection methods. But, it is very difficult to distinguish green fruits from green leaves [11], and a robust and special training dataset is needed [12] [13] [14]. Among different types of methods, the Hough Circle Transformation method was successfully used in detecting fruits [15]. To analyze images, a convex object identification algorithm was used for quick fruit detection. However, this method failed to identify fruits in various complex situations, such as where the fruit and leaves overlapped [16]. Fruits can also be detected using a combination of texture and color analysis [17]. In addition to texture, geometric shape analysis can also be used to identify fruits in images [18]. In order to detect citrus fruits, researchers conducted texture, color and shape analysis simultaneously [19]. However, texture analysis fails to detect fruits in most cases, except for high-resolution close-up images.

3.2.2 Machine learning based fruit detection

The initial study on distinguishing fruit from green leaves was published in 1967 by Brown and Schertz [20]. Apples, lemons, strawberries, bananas, etc. were detected using the Linear Classifier and the K-Nearest Neighbor Classifier [21] [22]. Later, K-Means Clustering was used for detection [23]. Since datasets were scarce at the start of the machine learning era, the accuracy of fruit detection was very poor. As a result, it was difficult to apply robot harvesting and fruit identification algorithms in real life. Later, AlexNet, a convolutional neural network was presented, and then various fruit detection studies and research work were introduced [24].

A new algorithm was proposed in 2016 named YOLO (You Only Look Once), which was a state-of-the-art real-time object detection system capable of easily detecting different types of objects from images [25]. In 2017, an algorithm based on instance segmentation, Mask R-CNN, was introduced, which was more accurate and reliable than any previous algorithms [26]. In 2018, YOLOv3, the third version of the YOLO algorithm, was introduced by Joseph Redmon and Ali Farhadi [27]. Liu et al. (2019) introduced a paper that used Mask R-CNN to detect green cucumber from green leaves [8]. Wang et al. (2020) presented a paper where waxberry fruit detection was performed using Mask R-CNN [10]. Currently, a new version of the YOLO algorithm, YOLOv5, has been released, which is used in this work [28].

3.3 MATERIALS AND METHODS

3.3.1 Image data acquisition

Images for the datasets were captured from a number of commercial lemon orchards. The images were collected using Huawei Y6 II thirteen Megapixel and Samsung galaxy S7 edge twelve Megapixel smartphone cameras and the resolution of the captured images were 4160 × 3120 pixels and 4032 × 3024 pixels, respectively. The size of the images varied from one megapixel to seven megapixels. Lighting, angles, and other relevant variables were meticulously monitored. Two samples of the collected images are shown in Figure 3.1.

Figure 3.1 Samples of collected images. Left, single lemon in bright light. Right, two lemons in low light.

3.3.2 Image pre-processing

To optimize the datasets, images were resized to the resolution of 1024 × 1024 pixels for Mask R-CNN and 416 × 416 pixels for YOLOv5 using photoshop. Data augmentation techniques such as crop and rotation were used to increase the datasets.

Figure 3.2 Sample of annotated images. Left, using VGG image annotator for Mask R-CNN. Right, using Make-Sense annotator for YOLOv5.

As shown in Figure 3.2, for Mask R-CNN, images of the dataset were annotated using the VIA (VGG image annotator) tool [29]. During annotation for Mask R-CNN, a polygon bounding box was used and saved as a JSON file. To annotate images for YOLOv5, Piotr Skalski's Make-Sense square bounding box tool was used and saved as a YOLO format file for YOLOv5 [30].

Table 3.1 Dataset configuration

Datasets	Methods	Annotation Tools	No. of Total Images	No. of Training Images	No. of Validation Images	No. of Test Images
Dataset 1	Mask R-CNN	VGG Image Annotator	700	500	100	100
Dataset 2	YOLOv5	Make-Sense Annotator	700	500	100	100

Table 3.1 shows the dataset configuration.

3.3.3 Model architecture

Mask R-CNN is a deep neural network that was developed as a solution to the instance segmentation problem, which can distinguish between various objects in an image or video [26]. For an input of a lemon image, the Mask R-CNN model returns bounding boxes, classes, and masks as shown in Figure 3.3.

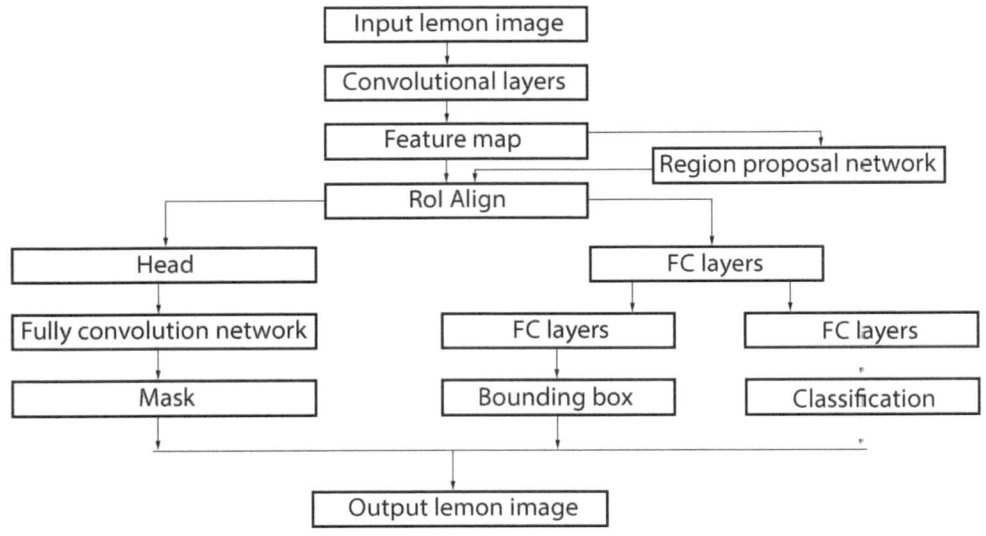

Figure 3.3 The structure of the Mask R-CNN model.

Glenn Jocher released YOLOv5 on June 26, 2020 [28]. It is the most recent version of YOLO (You Only Look Once), which is a real-time object detection model series. With the help of 58 open-source contributors, YOLOv5 has set a new standard for object detection models. It already outperformed the EfficientDet and previous YOLO versions [31].

3.3.4   Model training

The Mask R-CNN model used in this study has a feature pyramid network, and as the backbone, a residual neural network with 101 layers (ResNet-101) was used. This model was trained by dataset 1 mentioned in Table 3.1 with default network parameters. For training the model, 500 images with 1024×1024 resolution were used as a training set and 100 images as a validation set. In addition, 100 images were stored separately for the test set. The Google Colaboratory Tesla T4 graphics processing unit was used to train the model. The learning rate, batch size, and confidence were set to 0.001, 8, and 0.9, respectively.

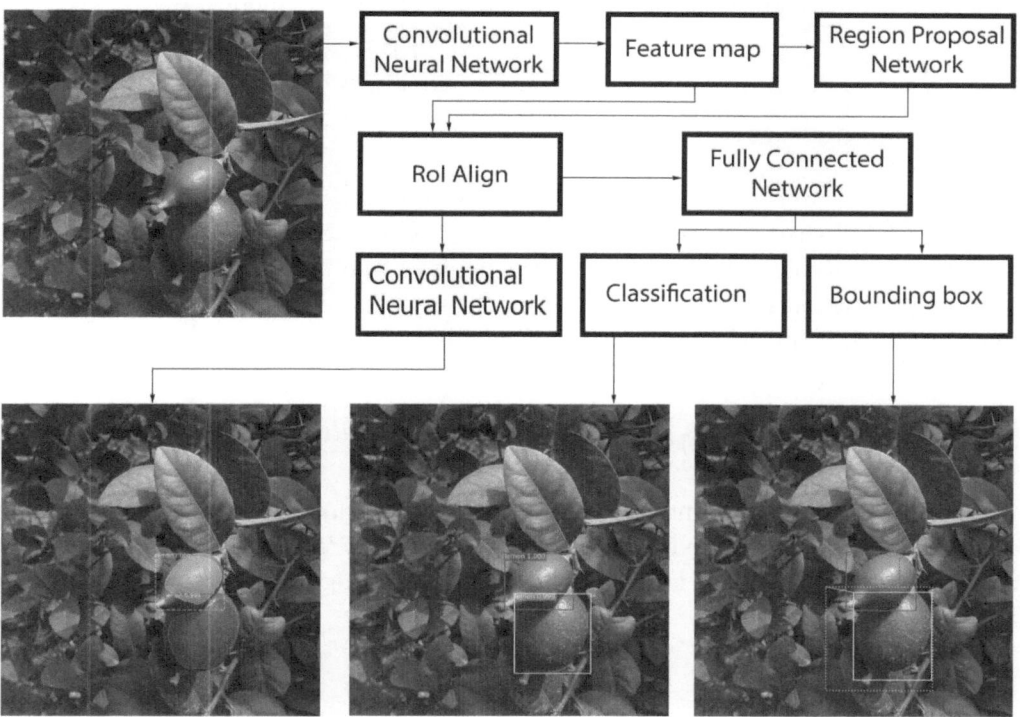

Figure 3.4   The workflow of the Mask R-CNN model.

The images from dataset 1 were processed by the pipeline illustrated in Figure 3.4. First, the pre-trained weight of the model was obtained from the COCO dataset. The feature extraction process for input images was carried out using the convolution layers. The feature map acquired from the convolutional layers was then transferred to the region proposal network. Regions of Interest (RoIs) were then generated from the region proposal network. The features extracted from each RoI were chosen by the RoI Align layers followed by passing the selected features to the convolution neural network and fully connected layers. Finally, classification, masking, and bounding-boxes occurred and were merged as output.

For training the YOLOv5 model, 500 images with a resolution of 416 × 416 were used as a training set and 100 images as a validation set. For the test set, 100 images were stored separately. The whole training process was carried out in Google Colaboratory with PyTorch 1.9.0+cu102 (Tesla K80). This model was trained on dataset 2 mentioned in Table 3.1 with default network parameters. Pre-trained YOLOv5s weight was used for trained weight generation. Batch size was set to 4 and confidence was set to 0.25.

3.4   RESULT ANALYSIS AND COMPARISON

3.4.1   Result analysis

For calculation, the variables Precision (3.1), Recall (3.2), F1 (3.3), FPR (3.4) and FNR (3.5) are used to measure the performance of the two models.

$$Precision(P) = \frac{TP}{TP + FP} \tag{3.1}$$

$$Recall(R) = \frac{TP}{TP + FN} \tag{3.2}$$

TP is true positive, FP is false positive, and FN is false negative. TP calculates the number of lemons detected correctly. FP calculates the number of objects that were incorrectly detected as lemons. FN is the false negative, which counts the number of lemons that were not detected. 'TP+FP' calculates all the objects that were detected as lemon and 'TP+FN' counts all the lemons present in the images.

$$F1 = \frac{2 \times Precision \times Recall}{Precision + Recall} \tag{3.3}$$

The variable F1 represents the capability or the performance of the models. F1 is measured using the weighted average of two variables: Precision (P) and Recall (R). A good fruit detection model should demonstrate a high precision value when the value of recall increases.

$$FPR = \frac{FP}{TP + FP} \tag{3.4}$$

$$FNR = \frac{FN}{TP + FN} \tag{3.5}$$

The FPR (False Positive Rate) is defined as the percentage of lemons detected incorrectly and FNR (False Negative Rate) is defined as the percentage of lemons that were undetected.

One example of sample detection by Mask R-CNN is shown in Figure 3.5.

Table 3.2 summarizes the performance analysis of the Mask R-CNN model. To validate this model, 100 images were tested from the test set. The model detected

Figure 3.5 Samples of detected lemons using the Mask R-CNN model. Left, multiple masked lemons. Right, single masked lemon.

Table 3.2 Lemon detection quality metrics of Mask R-CNN

Number of Test Images	Number of Lemons				P (%)	R (%)	F1 (%)	FPR (%)	FNR (%)
	Total	Detected	Not Detected	Incorrectly Detected					
100	253	213	40	31	87.2	84.1	85.6	12.7	15.8

213 lemons out of 253 lemons, while 40 remained undetected and 31 were detected incorrectly. The F1 score obtained from the Mask R-CNN model was 85.6%, which is comparable to the literature.

Table 3.3 Lemon detection quality metrics of YOLOv5

Number of Test Images	Number of Lemons				P (%)	R (%)	F1 (%)	FPR (%)	FNR (%)
	Total	Detected	Not Detected	Incorrectly Detected					
100	253	229	24	18	92.7	90.5	91.5	7.2	9.4

Table 3.3 summarizes the performance data of the YOLOv5 model. To validate this model, the test set of 100 images were used. The model detected 229 lemons out of 253 lemons, while 24 remained undetected and 18 were detected incorrectly. The F1 score obtained from the YOLOv5 model was 91.5%.

Figure 3.6 represents the detection of lemons enclosed in red rectangles by the YOLOv5 model.

3.4.2 Discussion

Both the Mask R-CNN and YOLOv5 models were trained adequately with a training set of 500 images and a validation set of 100 images each. The YOLOv5 model and the Mask R-CNN model can be used in harvesting robots in order to detect lemons

Figure 3.6 Samples of detected lemons using the YOLOv5 model. Left, pair of detected lemons. Right, single detected lemon.

in the orchards. This study can benefit commercial lemon producers and farmers in estimating the production, soil management, fertilizer use, and crop nutrition.

Table 3.4 Comparison of two methods Mask R-CNN and YOLOv5

Methods	Precision (%)	Recall (%)	F1 (%)	FPR (%)	FNR (%)
Mask R-CNN	87.2	84.1	85.6	12.7	15.8
YOLOv5	92.7	90.5	91.5	7.2	9.4

Table 3.4 summarizes the comparative analysis of both Mask R-CNN and YOLOv5 models used in detecting lemon fruits. The YOLOv5 model had superior performance in terms of F1 score, Recall and Precision by 6.8%, 7.6% and 6.3% respectively compared to the Mask R-CNN model. The performance of YOLOv5 is more accurate than Mask R-CNN because the model is very recent and less complex to train and test with PyTorch implementation. In addition, the YOLOv5 model yields better results while detecting small-sized and blurred lemons compared to the Mask R-CNN model. Furthermore, the YOLOv5 model has an advantage over the Mask R-CNN model in terms of computational speed.

Both the models showed unsatisfactory results for the cases where leaves have a shape similar to a lemon, they overlap, or clusters of lemons are present in the image and the image is blurry. As a result, FPR and FNR for the Mask R-CNN and YOLOv5 were found to be 12.7% and 15.8%, and 7.2% and 9.4%, respectively. These values can be minimized by training the models with more samples. However, in terms of accuracy, computational speed, and error count, the YOLOv5 model demonstrated superior performance and hence could be considered as a model of choice for fruit detection techniques.

Table 3.5 Comparison to other fruit detection models

Studies	Fruits	No. of Images	Methods	F1 Scores (%)
Bargoti and Underwood 2017 [7]	Almond	100	Faster R-CNN	77.5
Kang and Chen 2020 [32]	Apple	560	YOLOv3	86.0
Kang and Chen 2020 [32]	Apple	560	Mask R-CNN	86.8
Liu et al. (2019) [8]	Cucumber	1226	Improved Mask R-CNN	89.5
Bargoti and Underwood 2017 [7]	Apple	112	Faster R-CNN	90.4
Bargoti and Underwood 2017 [7]	Mango	270	Faster R-CNN	90.8
This work	Lemon	100	Mask R-CNN	85.6
This work	Lemon	100	YOLOv5	91.5

Table 3.5 compares the results of this work with other groups found in the literature in detecting different kinds of fruits. In terms of F1 scores, results are comparable and the YOLOv5 model demonstrates the highest value among all.

3.5 CONCLUSION

In this study, two machine learning models, Mask R-CNN and YOLOv5, were used to detect lemon fruits. To train these models, two datasets were prepared by collecting images of local lemon trees of all possible sizes and shapes common to Bangladesh. The performance of these two models in detecting lemon fruits were compared and evaluated. In terms of accuracy measures, the YOLOv5 model demonstrated better performance than the Mask R-CNN model, and this model is fast, accurate, and economical in terms of saving manual labor and improving lemon estimation. The performance of these two models is comparable to contemporary fruit detection models. The key challenges of this work were data collection, image annotations, identification of green lemons from the green leaves, and the fruits hidden inside the bush. The result of this work is inspiring considering the key challenges. In the future, these models can be trained to improve accuracy and can be extended to other fields of agriculture.

Bibliography

[1] Marti, Nuria, Pedro Mena, Jose Antonio Canovas, Vicente Micol, and Domingo Saura. 2009. "Vitamin C and the Role of Citrus Juices as Functional Food". In *Natural Product Communications* 4 (5): 1934578X0900400.

[2] Sarker, Md Nazirul Islam, M. Islam Sanjit Chandra Barman, R. Islam, and Amitabh Shuva Chakma. 2017. "Role of Lemon (Citrus Limon) Production on Livelihoods of Rural People in Bangladesh". In *Journal of Agricultural Economics and Rural Development* 2 (1): 167–175.

[3] Bangladesh Citrus Fruit Production, 1961-2020. In *Knoema.Com.* Accessed April 24, 2021. https://knoema.com.

[4] Bangladesh Lemon Suppliers, Wholesale Prices, and Market Information. In *Tridge.Com.* Accessed April 24, 2021. https://www.tridge.com.

[5] Tu, Shuqin, Yueju Xue, Chan Zheng, Yu Qi, Hua Wan, and Liang Mao. 2018. "Detection of Passion Fruits and Maturity Classification Using Red-Green-Blue Depth Images". In *Biosystems Engineering* 175: 156–67.

[6] Liu, Xiaoyang, Dean Zhao, Weikuan Jia, Wei Ji, and Yueping Sun. 2019. "A Detection Method for Apple Fruits Based on Color and Shape Features". In *IEEE Access* 7: 67923–33.

[7] Bargoti, Suchet, and James Underwood. 2017. "Deep Fruit Detection in Orchards". In *2017 IEEE International Conference on Robotics and Automation (ICRA).* IEEE.

[8] Liu, Xiaoyang, Dean Zhao, Weikuan Jia, Wei Ji, Chengzhi Ruan, and Yueping Sun. 2019. "Cucumber Fruits Detection in Greenhouses Based on Instance Segmentation". In *IEEE Access* 7: 139635–42.

[9] Stein, Madeleine, Suchet Bargoti, and James Underwood. 2016. "Image Based Mango Fruit Detection, Localisation and Yield Estimation Using Multiple View Geometry". In *Sensors* (Basel, Switzerland) 16 (11): 1915.

[10] Wang, Yijie, Jidong Lv, Liming Xu, Yuwan Gu, Ling Zou, and Zhenghua Ma. 2020. "A Segmentation Method for Waxberry Image under Orchard Environment". In *Scientia Horticulturae* 266 (109309): 109309.

[11] Bulanon, D. M., T. F. Burks, and V. Alchanatis. 2009. "Image Fusion of Visible and Thermal Images for Fruit Detection". In *Biosystems Engineering* 103 (1): 12–22.

[12] Mao, Wenhua, Baoping Ji, Jicheng Zhan, Xiaochao Zhang, and Xiaoan Hu. 2009. "Apple Location Method for the Apple Harvesting Robot". In *2009 2nd International Congress on Image and Signal Processing.* IEEE.

[13] Bulanon, D. M., and T. Kataoka. 2010. "Fruit Detection System and an End Effector for Robotic Harvesting of Fuji Apples". In *Agricultural Engineering International: CIGR Journal* 12 (1).

[14] Wei, Xiangqin, Kun Jia, Jinhui Lan, Yuwei Li, Yiliang Zeng, and Chunmei Wang. 2014. "Automatic Method of Fruit Object Extraction under Complex Agricultural Background for Vision System of Fruit Picking Robot". In *Optik* 125 (19): 5684–89.

[15] Whittaker, D., G. E. Miles, O. R. Mitchell, and L. D. Gaultney. 1987. "Fruit Location in a Partially Occluded Image". In *Transactions of the ASAE* 30 (3): 591–0596.

[16] Kelman, Eliyahu (efim), and Raphael Linker. 2014. "Vision-Based Localisation of Mature Apples in Tree Images Using Convexity". In *Biosystems Engineering* 118: 174–85.

[17] Zhao, J., J. Tow, and J. Katupitiya. 2005. "On-Tree Fruit Recognition Using Texture Properties and Color Data". In *2005 IEEE/RSJ International Conference on Intelligent Robots and Systems*. IEEE.

[18] Rakun, J., D. Stajnko, and D. Zazula. 2011. "Detecting Fruits in Natural Scenes by Using Spatial-Frequency Based Texture Analysis and Multiview Geometry". In *Computers and Electronics in Agriculture* 76 (1): 80–88.

[19] Kurtulmus, Ferhat, Won Suk Lee, and Ali Vardar. 2011. "Green Citrus Detection Using 'Eigenfruit', Color and Circular Gabor Texture Features under Natural Outdoor Conditions". In *Computers and Electronics in Agriculture* 78 (2): 140–49.

[20] Brown, G. K., and C. E. Schertz. 1967. "Evaluating Shake Harvesting of Oranges for the Fresh Fruit Market". In *Transactions of the ASAE* 10 (5): 577–0578.

[21] Bulanon, D. M., T. Kataoka, H. Okamoto, and S. I. Hata. 2004. "Development of a Real-Time Machine Vision System for the Apple Harvesting Robot". In *SICE 2004 annual conference*, 1:595–598. IEEE.

[22] Seng, Woo Chaw, and Seyed Hadi Mirisaee. 2009. "A New Method for Fruits Recognition System". In *2009 International Conference on Electrical Engineering and Informatics*. IEEE.

[23] Wachs, J. P., H. I. Stern, T. Burks, and V. Alchanatis. 2010. "Low and High-Level Visual Feature-Based Apple Detection from Multi-Modal Images". In *Precision Agriculture* 11 (6): 717–35.

[24] Krizhevsky, Alex, Ilya Sutskever, and Geoffrey E. Hinton. 2017. "ImageNet Classification with Deep Convolutional Neural Networks". In *Communications of the ACM* 60 (6): 84–90.

[25] Redmon, Joseph, Santosh Divvala, Ross Girshick, and Ali Farhadi. 2016. "You Only Look Once: Unified, Real-Time Object Detection". In *2016 IEEE Conference on Computer Vision and Pattern Recognition (CVPR)*. IEEE.

[26] He, Kaiming, Georgia Gkioxari, Piotr Dollar, and Ross Girshick. 2017. "Mask R-CNN". In *2017 IEEE International Conference on Computer Vision (ICCV)*. IEEE.

[27] Redmon, Joseph, and Ali Farhadi. 2018. "YOLOv3: An Incremental Improvement". ArXiv [Cs.CV]. http://arxiv.org/abs/1804.02767.

[28] Jocher, Glenn. 2020. "Ultralytics/yolov5: YOLOv5 in PyTorch". In *GitHub*. https://github.com/ultralytics/yolov5.

[29] Dutta, Abhishek, and Andrew Zisserman. 2019. "The VIA Annotation Software for Images, Audio and Video". In *Proceedings of the 27th ACM International Conference on Multimedia*. New York, NY, USA: ACM.

[30] Skalski, Piotr. SkalskiP/Make-Sense: Free to Use Online Tool for Labelling Photos. 2019. Https://Makesense.ai. In *GitHub*. https://github.com/SkalskiP/make-sense.

[31] Kuznetsova, Anna, Tatiana Maleva, and Vladimir Soloviev. 2020. "Detecting Apples in Orchards Using YOLOv3 and YOLOv5 in General and Close-up Images". In *Advances in Neural Networks – ISNN 2020*, 233–43. Cham: Springer International Publishing.

[32] Kang, Hanwen, and Chao Chen. 2020. "Fruit Detection, Segmentation and 3D Visualisation of Environments in Apple Orchards". In *Computers and Electronics in Agriculture* 171: 105302. doi:10.1016/j.compag.2020.105302.

CHAPTER 4

A Deep Learning Approach in Detailed Fingerprint Identification

Mohiuddin Ahmed

Rajshahi University of Engineering & Technology, Rajshahi, Bangladesh

Abu Sayeed

Rajshahi University of Engineering & Technology, Rajshahi, Bangladesh

Azmain Yakin Srizon

Rajshahi University of Engineering & Technology, Rajshahi, Bangladesh

Md Rakibul Haque

Rajshahi University of Engineering & Technology, Rajshahi, Bangladesh

Md. Mehedi Hasan

Rajshahi University of Engineering & Technology, Rajshahi, Bangladesh

CONTENTS

Fingerprints, as one of the most acceptable biometrics, has a significant use for security purposes. In today's world, fingerprints are utilized in criminal investigations for identification purposes. Fingerprints can be used to identify a person's gender, hand, finger, and other relevant aspects. The required time and effort in identifying an individual can be reduced by gender, hand, and finger classifications using fingerprints. In this chapter, a very simple model based on Convolutional Neural Networks (CNNs) is proposed to classify fingerprints by gender, hand, and finger. This research is conducted on the publicly available Sokoto Coventry Fingerprint Dataset (SOCOFing). Our proposed model outperformed existing state-of-the-art approaches by a significant amount, with gender, hand, and finger accuracy rates of 96.50%, 97.83%, and 93.88%, respectively.

4.1 INTRODUCTION

A fingerprint is a unique mark of an individual and is considered the oldest and most acceptable method of personal identification [1]. Because of the ease of acquisition, preprocessing, storage of fingerprint images and the minimal effort required from the user, it has become one of the most acceptable biometrics worldwide. The lower price of the sensors required to acquire fingerprint images can be considered as another reason. Moreover, the fingerprint pattern remains unchanged throughout the whole life of a person if there is no skin disease or permanent scar. No two people have been found to have the same fingerprint. Even ten fingerprints from ten different fingers of the same person are different.

With the increased use of the fingerprint as a biometric identification system, the necessity for automatic identification of fingerprints has come to the fore [2]. However, fingerprint detection is a very daunting task because of the noise and rotation variance associated with the image. This involves the detection of intricate features from fingerprint image [3]. Furthermore, with the ever-growing use of fingerprint identification, the number of samples is increasing drastically. That may lead to more computational cost while performing identification of an individual. Recently, Convolutional Neural Network has achieved significant performance in image classification due to its ability for extraction of prominent features from complex images [4]. CNN can extract complex features from fingerprint images and help in the process of automatic fingerprint identification and in turn, derivative characteristic details of an individual. Deriving characteristic details from a fingerprint can reduce the search space by a lot while performing automatic fingerprint detection for an individual. In this chapter, a model based on CNNs is proposed to classify gender, hand, and finger from fingerprints. The main contributions of this work can be outlined as follows:

- A very simple baseline convolutional neural network model with a small number of parameters is proposed that works both as a feature extractor and a classifier.

- This study includes a comparison to comparable state-of-the-art works. Our suggested model outscored previous research by a significant margin.

The rest of this chapter is organized as follows. Related works are described in Section 4.2. A brief description of the dataset is given in Section 4.3. In Section 4.4, the proposed methodology is introduced. The experimental setup and implementation are described in Section 4.5. Section 4.6 shows the results and discussion. Finally, in Section 4.7, we conclude our work.

4.2 RELATED WORKS

Many researchers all over the world have studied fingerprint-based classification using different techniques with a view to achieving different types of research objectives.

Shehu et al. [5] applied a transfer learning approach and provided a benchmark result of the publicly available SOCOFing dataset with a classification accuracy of 75.2% for gender, 93.5% for hand, and 76.72% for fingers. Giudice et al. [6] applied Inception-v3 architecture for training and proposed a method for detecting fingerprint alteration, alteration types, and classifying gender, hand, and fingers. They also produced activation maps showing the area of a fingerprint focused by a neural network model to learn. The main contribution of this paper was to provide a single architecture to solve all classification problems related to fingerprints. Shehu et al. [7] proposed a CNN model to identify different types of fingerprint alterations and achieved an accuracy of 98.55%. They also provided a comparison with a pre-trained residual CNN architecture that produced an accuracy of 99.88%. Sheetlani et al. [8] collected their own dataset of 750 fingerprints. They applied contrast-limited adaptive histogram equalization for preprocessing, discrete wavelet transform for feature extraction, and a feed-forward back-propagation neural network for classification. They achieved an accuracy of 96.60%. Nithin et al. [9] worked on a dataset of fingerprints acquired from 550 subjects to identify the gender based on the ridge count of the fingerprint. They applied Bayes' theorem and found that the ridge count for a female is more than for a male.

4.3 DATASET

This study is conducted on the Sokoto Coventry Fingerprint Dataset (SOCOFing) containing 6000 fingerprints from 600 African people aged 18 years or older. The SOCOFing dataset contains information like gender, hand and finger name. Moreover, synthetically altered images with three different alteration levels – easy, medium, and hard – provided in this dataset. Fig. 10.1 shows the images of real and altered versions of fingerprints. There are three types of alterations – obliteration, central rotation, and Z-cut. Examples of different types of alterations are shown in Fig. 6.2. The STRANGE toolbox was used to perform these alterations over the images with a resolution of 500dbi [11]. In total, there are 55270 images, including both real and altered versions of fingerprints, of dimension $103 \times 96 \times 1$ (height, width, gray) in this dataset.

4.4 METHODOLOGY

A model based on Convolutional Neural Networks is developed for gender, hand and finger classification. Fig. 6.3 shows the architecture of our proposed methodology. The input images are first preprocessed and then fed into the proposed CNN model. The

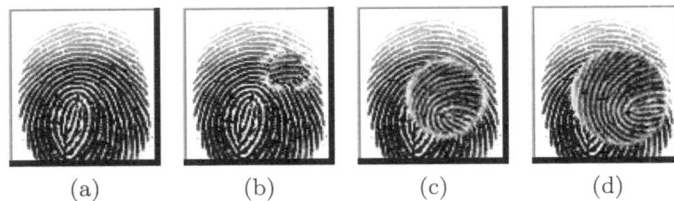

Figure 4.1 Images of the real and altered versions of fingerprints: (a) Real Fingerprint, (b) Altered (Easy) Fingerprint, (c) Altered (Medium) Fingerprint, (d) Altered (Hard) Fingerprint.

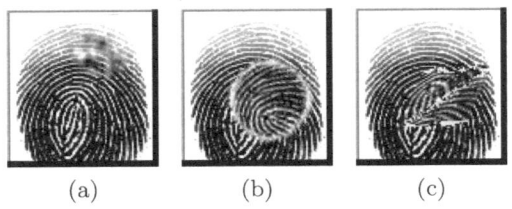

Figure 4.2 Different types of alterations: (a) Obliteration, (b) Central Rotation and (c) Z-cut.

model is composed of stacks that include three convolution layers, three max-pooling layers, fully connected layers, and the softmax activation function at the end.

4.4.1 Convolutional Neural Network Model

The Convolutional Neural Network is a deep learning algorithm designed to learn spatial features adaptively. CNN consists of convolutional layers, maxpooling layers, and fully connected layers. Because of the capability of adaptive learning, CNN is broadly used in classification of images.

We propose a very simple baseline CNN model consisting of three convolutional layers with 32, 3 × 3 filter kernels for feature extraction. Same padding was used in all three cases. Rectifier linear unit (ReLU) was applied as the activation function for the output of the convolutional layers. Three max-pooling layers were applied for dimensionality reduction. The final output of the convolutional layers was flattened using the flatten layer and passed to the fully connected layers consisting of two dense layers and the softmax activation function at the end.

4.5 EXPERIMENTAL SETUP AND IMPLEMENTATION

The SOCOFing dataset contains 55270 fingerprints of dimension $103 \times 96 \times 1$ (height, width, gray). First, the images were resized to the dimension of $96 \times 96 \times 1$. 50% for training, 20% for validation, and the remaining 30% for testing. The resized images of the training set were fed into the CNN model.

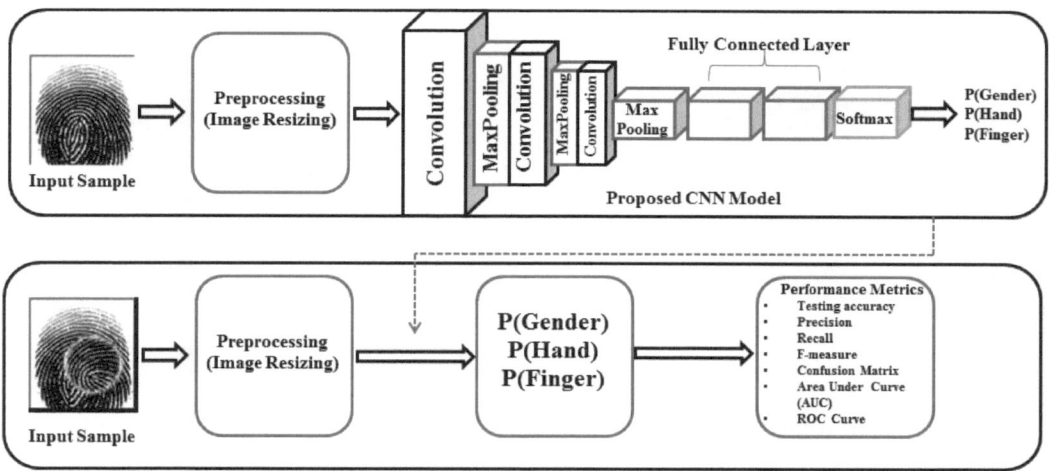

Figure 4.3 Architecture of the proposed methodology.

4.5.1 Hyperparameter Optimization

All the hyperparameters were optimized using the grid search technique. The Adam optimizer was used to train the model with a learning rate of 0.001. In the case of a plateau, after every 5 epochs, if the validation accuracy was not increased, the learning rate was reduced by a factor of 0.2. The minimum learning rate was considered 0.0000001. Cross entropy was used as the loss function. A total of 100 epochs were performed with the batch size of 128 and the best weight vector was saved. After training, the weight vector was used to classify gender, hand, and finger.

4.5.2 Evaluation Criteria

Various performance metrics, including training and validation accuracy, loss curve, ROC curve, confusion matrix, accuracy, precision, recall, F-measure, and Area Under Curve (AUC) were used to evaluate the results obtained from experimental analysis. In the case of finger classification, the micro average technique was used to compute precision, recall, and F-measure.

4.6 RESULTS AND DISCUSSION

In this chapter, we considered gender, hand, and finger classifications.

4.6.1 Gender Classification

Gender classification, based on 44375 male and 10895 female samples, is a binary classification problem. It is clear that the dataset is class-imbalanced. As other target attributes (hand and finger) are not class-imbalanced, no technique was applied to solve the class-imbalance problem. The testing accuracy of gender classification is 96.50%. The Area Under Curve (AUC) is 0.9344. Fig. 6.4 shows the training and validation loss and accuracy curves and the ROC curve for gender. The confusion

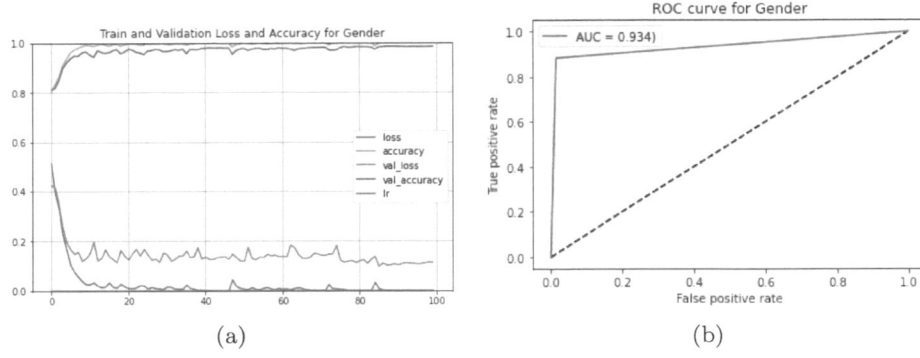

(a) (b)

Figure 4.4 (a) Training and validation loss and accuracy curves and (b) ROC curve for gender.

Table 4.1 Confusion matrix for gender.

Actual Label	Predicted Label	
	Male	Female
Male	13085	197
Female	384	2915

matrix is depicted in Table 4.1. From the confusion matrix, we can see that 16,000 out of 16,581 test samples are classified correctly.

4.6.2 Hand Classification

Hand classification, containing 27806 left-hand and 27464 right-hand samples, is a binary classification problem. It is clear that the dataset is quite class-balanced. The testing accuracy of hand classification is 97.83%. The Area Under Curve (AUC) is 0.9783. Fig. 4.5 shows the training and validation loss and accuracy curves and the ROC curve for the hand. The confusion matrix is depicted in Table 4.2. From the confusion matrix, we can see that 16,221 out of 16,581 test samples are classified correctly.

4.6.3 Finger Classification

Finger classification is a multi-classification problem. From Fig. 4.6(a), it is clear that the dataset is quite class-balanced. The testing accuracy of finger classification

Table 4.2 Confusion matrix for hands.

Actual Label	Predicted Label	
	Left	Right
Left	8191	161
Right	199	8030

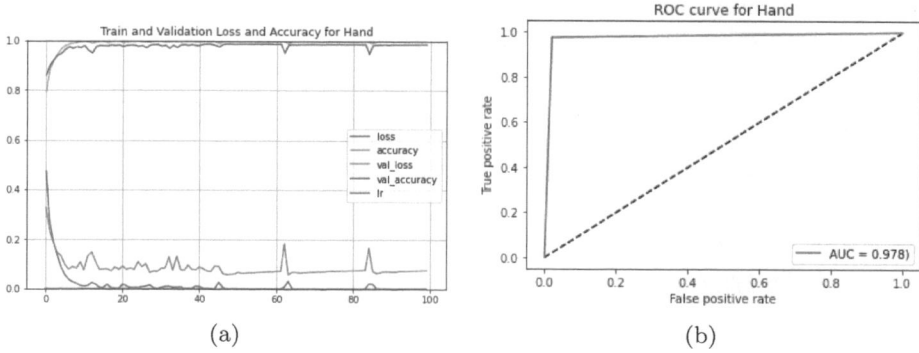

(a)   (b)

Figure 4.5 (a) Training and validation loss and accuracy curves and (b) ROC curve for hand.

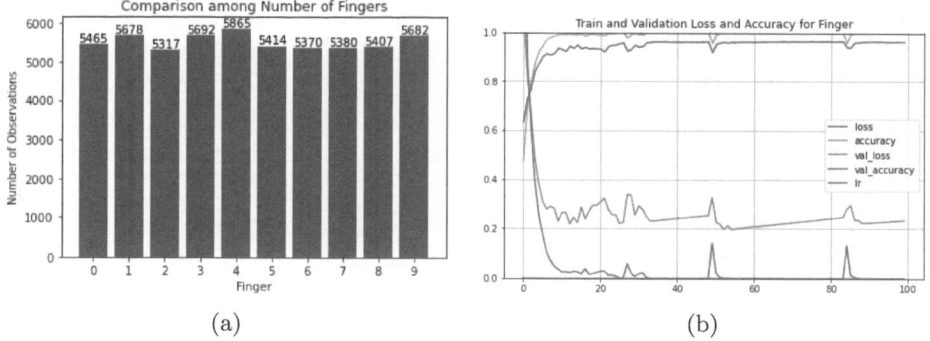

(a)   (b)

Figure 4.6 (a) Number of different fingers in the finger class and (b) training and validation loss and accuracy curves for curves.

Table 4.3 Confusion matrix for fingers.

Actual Label	Predicted Label									
	0	1	2	3	4	5	6	7	8	9
0	1539	31	22	14	0	19	0	5	6	0
1	16	1644	22	10	5	2	23	11	8	0
2	22	34	1399	51	4	1	11	11	3	0
3	14	7	38	1519	22	44	6	14	10	3
4	8	0	6	10	1504	3	3	0	0	1
5	28	5	20	26	1	1533	35	13	13	4
6	12	9	4	15	2	25	1601	23	7	0
7	3	13	7	4	3	12	69	1544	15	12
8	6	2	0	7	0	4	0	43	1652	6
9	1	0	4	2	2	6	6	2	23	1632

Table 4.4 Evaluation results of gender, hand, and finger classifications.

Classification	Accuracy(%)	Precision(%)	Recall(%)	F-measure(%)	AUC
Gender	96.50	97.15	98.52	97.83	0.9344
Hand	97.83	97.63	98.07	97.85	0.9783
Finger	93.88	93.28	93.78	93.87	-

Table 4.5 Comparison of accuracies in gender, hand, and finger classification with other state-of-the-art works.

Works	Gender	Hand	Finger
Shehu et al. [5]	75.20%	93.50%	76.72%
Giudice et al. [6]	92.52%	97.53%	92.18%
Our Model	**96.50%**	**97.83%**	**93.88%**

is 93.88%. Fig. 4.6 shows the comparison of the number of fingers and the training and validation loss and accuracy curves for fingers. The confusion matrix is depicted in Table 4.3. Here 0 to 4 and 5 to 9 indicate fingers of the left and right hands, respectively. The 0 and 5 indicate little fingers. Similarly, 1 and 6, 2 and 7, 3 and 8, 4 and 9 indicate the ring, middle, and index fingers and thumb of the left and right hand, respectively.

Because it is a very simple baseline architecture with a small number of parameters, our proposed model performed well in all three classifications. Training from scratch with the simple architecture worked out well in this case because of the simple nature of the dataset. The performance of this proposed model is shown in Table 4.4. Our proposed model outperformed the recent works in all three classification by a significant amount. Comparisons of accuracies in gender, hand, and finger classification with other state-of-the-art works are depicted in Table 4.5. From this table, we clearly observe that the accuracy rate of our proposed model for gender, hand, and finger classifications are higher than that of Shehu et al. [5] and Giudice et al. [6].

4.7 CONCLUSION

Fingerprints are unique human attributes. It is considered a shred of prime evidence in criminal investigation and identification of individuals. In this chapter, we proposed a very simple baseline model based on CNNs to classify gender, hand, and finger from fingerprints. This model has a very small number of parameters compared to other complex models used in previous works. Our proposed model achieved an accuracy of 96.50%, 97.83%, and 93.88% for gender, hand, and finger, respectively. We also provided a comparison with other state-of-the-art works. Because of the dataset's simplicity, training from scratch with the basic architecture performed well compared to other recent works in this field.

In the future, other aspects of the dataset, like fakeness detection and alteration-type detection will be considered for classificaton. We will try to improve the robustness of our model. This model will be applied on other publicly available datasets also.

Bibliography

[1] Jain, A. K., Bolle, R., & Pankanti, S. (Eds.). (2006). *Biometrics: Personal Identification in Networked Society* (Vol. 479). Springer Science & Business Media.

[2] Uchida, K. (2005). Fingerprint Identification. *NEC Journal of Advanced Technology, 2*(1), 19-27.

[3] Nogueira, R. F., de Alencar Lotufo, R., & Machado, R. C. (2014, October). Evaluating Software-based Fingerprint Liveness Detection Using Convolutional Networks and Local Binary Patterns. In *2014 IEEE Workshop on Biometric Measurements and Systems for Security and Medical Applications (BIOMS) Proceedings* (pp. 22-29). IEEE.

[4] Krizhevsky, A., Sutskever, I., & Hinton, G. E. (2012). Imagenet Classification with Deep Convolutional Neural Networks. *Advances in Neural Information Processing Systems, 25,* 1097-1105.

[5] Shehu, Y. I., Ruiz-Garcia, A., Palade, V., & James, A. (2018, December). Detailed Identification of Fingerprints Using Convolutional Neural Networks. In *2018 17th IEEE International Conference on Machine Learning and Applications (ICMLA)* (pp. 1161-1165). IEEE.

[6] Giudice, O., Litrico, M., & Battiato, S. (2020, October). Single Architecture and Multiple Task Deep Neural Network for Altered Fingerprint Analysis. In *2020 IEEE International Conference on Image Processing (ICIP)* (pp. 813-817). IEEE.

[7] Shehu, Y. I., Ruiz-Garcia, A., Palade, V., & James, A. (2018, October). Detection of Fingerprint Alterations Using Deep Convolutional Neural Networks. In *International Conference on Artificial Neural Networks* (pp. 51-60). Springer, Cham.

[8] Sheetlani, J., & Pardeshi, R. (2017). Fingerprint Based Automatic Human Gender Identification. *Int. J. Comput. Appl, 170*(7), 1-4.

[9] Nithin, M. D., Manjunatha, B., Preethi, D. S., & Balaraj, B. M. (2011). Gender Differentiation by Finger Ridge Count among South Indian Population. *Journal of Forensic and Legal Medicine, 18*(2), 79-81.

[10] Shehu, Y. I., Ruiz-Garcia, A., Palade, V., & James, A. (2018). Sokoto Coventry Fingerprint Dataset. *arXiv preprint arXiv:1807.10609.*

[11] Papi, S., Ferrara, M., Maltoni, D., & Anthonioz, A. (2016, September). On the Generation of Synthetic Fingerprint Alterations. In *2016 International Conference of the Biometrics Special Interest Group (BIOSIG)* (pp. 1-6). IEEE.

CHAPTER 5

Probing Skin Lesions and Performing Classification of Skin Cancer Using EfficientNet while Resolving Class Imbalance Using SMOTE

Md Rakibul Haque

Rajshahi University of Engineering & Technology, Rajshahi, Bangladesh

Azmain Yakin Srizon

Rajshahi University of Engineering & Technology, Rajshahi, Bangladesh

Mohiuddin Ahmed

Rajshahi University of Engineering & Technology, Rajshahi, Bangladesh

CONTENTS

With the ever-increasing radiation, skin cancer is becoming one of the most common cancers found in people of all races. Earlier, skin cancer was diagnosed through manual screening and dermoscopic analysis by an expert dermatologist. However, this process is quite costly and time-consuming. These limitations and the recent breakthrough in deep learning architectures in medical imaging inspired us to use these architectures for analyzing skin lesions. Some skin cancers are more common than other types of cancer that create class imbalance. In this chapter, we have proposed a new approach where we have used SMOTE for oversampling and then an EfficientNetB0 architecture has been used to extract prominent features from complex skin lesion images. We have used HAM10000, a benchmark dataset containing images of seven distinct classes of skin cancers, for evaluation. We have achieved an accuracy of 96.56%, which outperforms all the other benchmark approaches by a significant margin.

5.1 INTRODUCTION

Skin cancer mainly refers to the abnormal growth of skin cells. Skin cancers are preliminarily classified according to the position of the affected cells in three layers. In general, skin cancers can be classified into two categories, one is Non-Melanoma cancer (NMC), which occurs in the outermost layer or the middle layer of skin, and another one is Melanoma cancer (MC), which takes place in the innermost layer of skin [1]. NMC has a low mortality rate and contagion rate among skin tissue. Now Malignant Melanoma is one of the most lethal skin cancers of all time. It has a high mortality rate as it accounts for 75% of deaths caused by skin cancers. If not detected early, melanoma cancer spreads to other parts of the body and renders any treatment moot. The number of patients with skin cancer reached a total of 1,329,781, which was the third-highest among all the cancers in that year [2]. Detection and diagnosis of skin cancer at an early stage with precise accuracy is an important task to tackle the ever-increasing skin cancer epidemic.

In the past, dermoscopy was being used for detecting skin cancers. But this approach requires lots of expertise as the dermatologists need to differentiate among an enormous number of skin lesions [3]. The complex nature of skin lesions in the image makes dermoscopy a cumbersome task [4]. This approach is highly prone to subjective judgment, which reduced the efficacy. Despite introducing many algorithms and scoring systems [5], such as the ABCD-E rule, seven Score System to improve this clinical diagnosis, or dermoscopy, the performance remains subpar. This has influenced researchers to introduce automation and use different kinds of computational approaches to perform proper diagnosis and detection of skin cancer.

Recently, deep learning architecture has achieved remarkable success in the diagnosis of skin cancers [6]. Deep Convolutional Neural Networks possess the ability to process general and highly variable tasks in complex and detailed images [7]. This specific ability enables deep CNN to extract more prominent features than handcrafted features from skin lesion images. Mohamed used MobileNet and DenseNet for the classification of skin cancer and achieved an accuracy of 92.7% and 91.2%, respectively [8]. However, he used a simple augmentation process to bring balance into the dataset. Al-Masni [9] evaluated the performance of four benchmark CNN architectures: Inception-v3, ResNet-50, Inception-ResNet-v2, and DenseNet-201 for classification types of skin cancer from the HAM10000 dataset and achieved an accuracy of 88.05%, 89.28%, 87.74%, and 88.70%, respectively for the architectures mentioned above. The problems with the above work are that they have used models with a huge number of parameters and haven't accounted for the class imbalance problem. EfficientNet [10] is one of those benchmark CNN architectures that has achieved remarkable accuracy in the medical imaging field for its significant feature extraction from the raw input image.

Along with feature extraction, another problem with skin cancer classification is class imbalance [11]. Some skin cancers are more frequent than others. This creates class imbalance while performing skin cancer classification. Oversampling increases the minority class samples and makes them the same as majority class samples. Oversampling is mainly done by repetition or synthetic approaches such as the Synthetic Minority Oversampling Technique (SMOTE) [12]. SMOTE produces new instances based on interpolation. This results in a much more suitable sample contrast than just imitation or random oversampling.

In this chapter, we have proposed a novel approach that uses SMOTE for oversampling. After that, it uses a transfer learned EfficientNet architecture specifically tailored for skin cancer detection to extract complex and finesse features from skin lesions for better diagnosis and classification. We have achieved results that are superior to other benchmark methods on the HAM10000 dataset. The main contribution of this research work is that it has tailored an EfficientNet specifically for analyzing skin lesions while removing imbalance from the dataset.

The rest of the chapter is structured as follows. In Section 5.2, a detailed discussion of the dataset, SMOTE, and EfficientNet are presented. Section 5.3, provides a brief discussion of the proposed approach. In Section 5.4, experimental setup and classification results are described in detail. Finally, a conclusion is drawn in Section 5.5.

5.2 METHODOLOGY

5.2.1 Dataset Description

For proper evaluation of a proposed approach, the first and foremost thing is a benchmark dataset. In the early days, there was a lack of images of skin lesions and varieties of those images. These two factors hampered the training process of deep neural networks. To tackle this problem, International Skin Imaging Collaboration (ISIC) introduced HAM10000 dataset [13] in 2018. This dataset consists of 10,015 dermoscopic images belonging to seven different classes of skin cancer, which represent a collection of all the important diagnostic groups in the domain of skin cancers. Actinic keratoses (akiec), basal cell carcinoma (bcc), benign keratosis-like lesions (bkl), dermatofibroma (df), melanoma (mel), melanocytic nevi (nv), and vascular lesions (vasc) are the seven different classes. More than half of the ground truth for skin lesion images is determined using histopathology. The rest of them are confirmed using

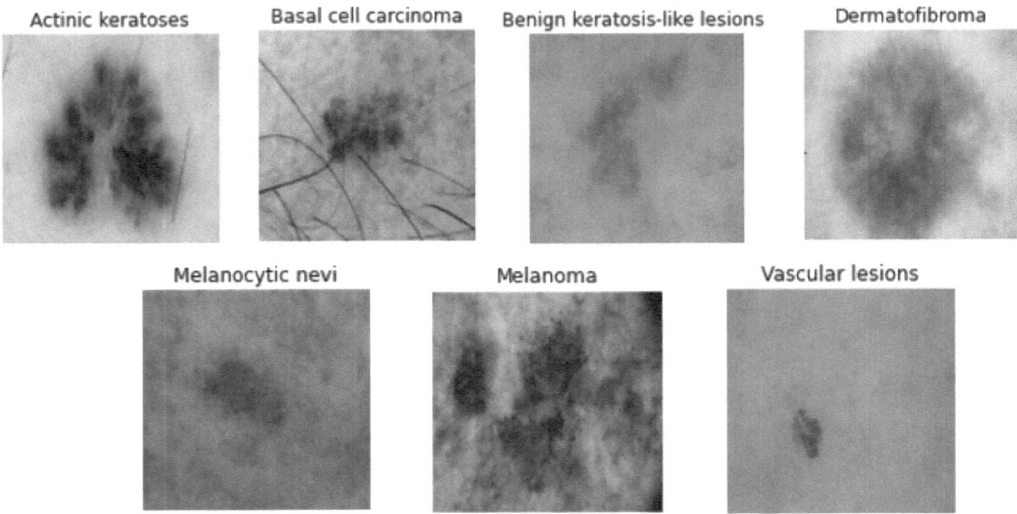

Figure 5.1 Graphical representation of different types of skin cancer

follow-up examination, expert consensus, or Vivo confocal microscopy. The dataset is immensely imbalanced as the majority class, melanocytic nevi consists of 6,705 images, whereas the minority class vascular lesions consist of only 115 images. Figure 5.1, displays images for different types of skin cancer. Figure 5.2 represents the distribution of different classes that indicates class imbalance. We have performed a 70:30 split for training and test samples. Furthermore, 20% of training data has been used as validation samples.

5.2.2 SMOTE

SMOTE is referred to as a benchmark oversampling technique for resolving class imbalance [14]. Instead of simply copying or imitating the value of minority class samples, smote focuses on creating synthetic samples based on different types of interpolation. In smote, new samples are created by observing feature space, the value of features, and the relationship between the feature values. That is the reason why smote is called a feature space-based technique. Let's consider X_1, a minority class sample, that is the base for creating synthetic samples for that specific minority class. After that, based on a distance metric, several random points are chosen from that class $(X_{11}, X_{12}, X_{13} and X_{14})$. After that, interpolation is performed to create the new synthetic sample r_1 as portrayed in Figure 5.3.

The brief working procedure of smote is described as follows. At first, the total amount of oversampling, N, for a minority class is determined. This number is preliminarily chosen in such a way that it ensures a similar distribution of major and minor groups in the dataset. After that, a repetitive process is performed consisting of several steps. Firstly, a sample is selected randomly from the minority class in which we want to perform oversampling. Secondly, the k number of neighbor instances of

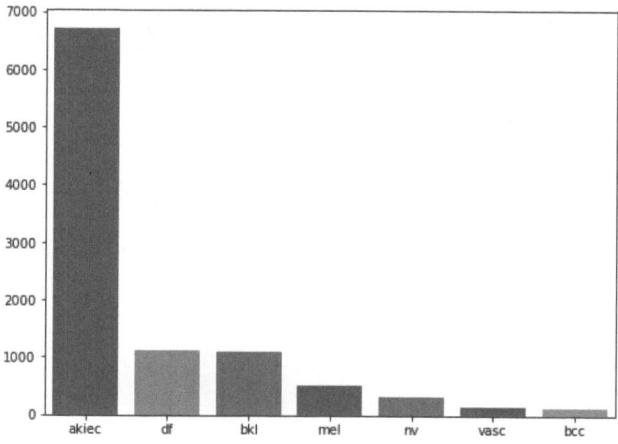

Figure 5.2 Distribution of image samples across seven classes

that sample is chosen according to a distance-based metric. Finally, N of these k neighbors is used to re-create the synthetic samples by interpolation. To create new samples, the difference between the original feature vector (samples) and its neighbor is taken and multiplied by a random number between 0 to 1 and added to the previous feature vector (neighbor). This creates a new sample along the line segment between the existing feature vectors or samples [14]. The same procedure is followed while oversampling skin lesion images.

5.2.3 EfficientNet

As model scaling does not modify layer operators in the baseline network, producing a better baseline network is also complex. To adequately illustrate the effectiveness of the scaling method, a new mobile-size baseline, called EfficientNet, was constructed [10]. Inspired by [15], it was introduced by a baseline network by leveraging a multi-objective neural structure search that optimizes both accuracy and FLOPS. Therefore, the authors optimized FLOPS despite latency as they were not targeting any definite hardware design. The search provided an efficient network, which was named EfficientNetB0. Since the identical search area was practiced as [15], the design is comparable to MnasNet, although EfficientNetB0 is somewhat bigger because of the larger FLOPS target of 400M. Beginning from the baseline EfficientNet-B0, they employed the compound scaling process to scale it up. They set various constants and scaled up the baseline network to achieve EfficientNetB1 to EfficientNetB7. The EfficientNet architecture can be represented mathematically as:

$$N(d, w, r) = \underset{i=1..s}{\odot} \hat{F}_i^{d.L^i}(X_{\langle r.\hat{H}_i.r.\hat{W}_i.w.\hat{C}_i\rangle}) \tag{5.1}$$

Here, N is the network and d, w, r are coefficients for scaling the network along width, depth and resolution, $\hat{F}_i$ denotes the convolution operation for the ith stage, and L_i represents the number of times $\hat{F}_i$ is repeated. $\hat{H}_i, \hat{W}_i$ and $\hat{C}_i$ simply denote

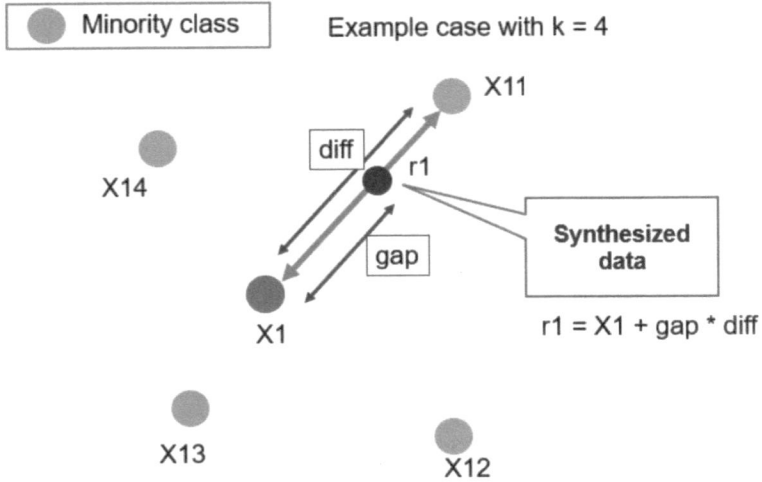

Figure 5.3 Graphical representation of the basic working principle of SMOTE

the input tensor shape for i^{th} stage. d, w, r can be further broken into the following form:

$$d = \alpha^{\phi}, w = \beta^{\phi}, r = \gamma^{\phi}, \alpha . \beta^2 . \gamma^2 \approx 2, \alpha, \beta, \gamma \geq 1 \qquad (5.2)$$

Here, ϕ is a global variable that determines the number of available resources, and α, β, and γ determine how the resources are allocated to network depth, width and resolution, respectively.

5.3 PROPOSED APPROACH

An overview of our proposed approach is illustrated in Figure 5.4. The key steps of our proposed approach will be discussed below in detail.

5.3.1 Resolving Class Imbalance Using SMOTE

As discussed previously, this dataset was highly imbalanced, which has caused a great barrier while training any feature extraction and classification model. To resolve this problem, we have used Smote in oversampling the skin lesion images. As smote works based on feature space rather than considering the data as a whole, Smote can create better synthetic samples in comparison to traditional augmentation methods. We have oversampled the training samples so that the classification model can learn about each sample in an unbiased manner. Although there are different kinds of augmentation techniques available, Smote was chosen because it is based on the feature space, not the data space.

5.3.2 Extracting Complex and Versatile Features Using EfficientNetB0

For extracting prominent and effective features from complex skin lesion images, we have used EfficientNetB0. Among seven types of EfficientNet architectures, we have

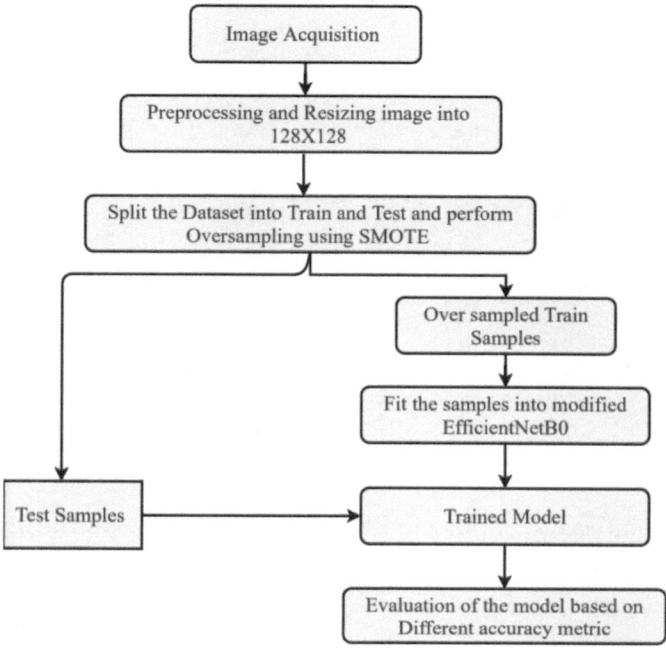

Figure 5.4 An overview of our proposed approach for skin cancer classification

used the B0 architecture because it has less computational cost than other Efficient-Net architectures. The EfficientNetB0 architecture was designed based on a neural network search that maximizes accuracy and minimizes floating-point operation. At first, we modified the architecture to accept input of size 128×128 because it is orig-inally trained with images of size 224×224. After that, we initialized the model with weights from the Imagenet challenge. This has drastically reduced computational time. Then, we trained the model using our training samples for tuning EfficientNet for skin classification. After that, we used a Multilayer Perceptron [16] with three hidden layers to perform accurate classification based on the superior features ex-tracted using EfficientNetB0. The number of hidden layers and nodes in each layer were calculated by grid search in a heuristic search space. Finally, the output layer consisted of seven nodes with a Softmax [17] activation function. Figure 5.5 illustrates a graphical representation of EfficientNetB0 that has been specially tailored for skin cancer classification.

5.4 EXPERIMENTAL ANALYSIS

5.4.1 Experimental Setup

We utilized Keras [18], a deep learning framework, for creating our EfficientNetB0 architecture, which is specifically modified for skin cancer classification. We trained the model for 60 epochs with a batch size of 128. We used variable learning rates throughout the training process starting from 0.00001 initially. Furthermore, we used

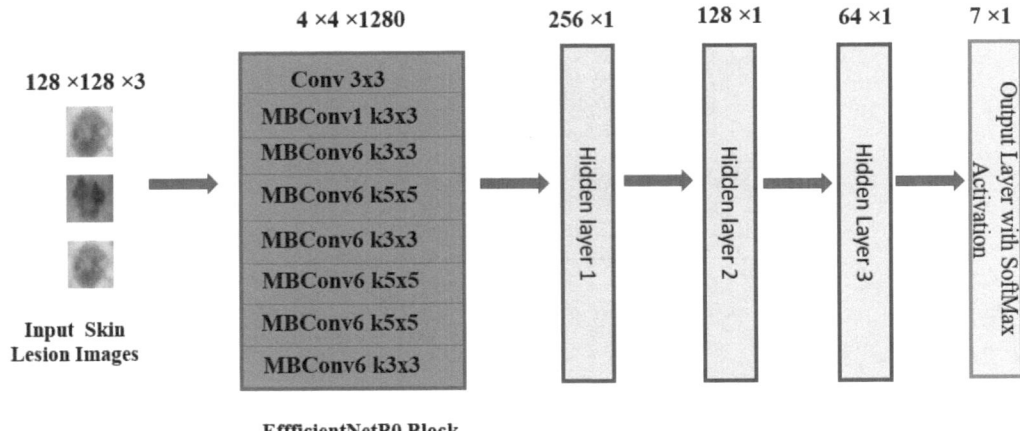

Figure 5.5 The architecture of our modified EfficientNetB0 specifically for performing skin cancer classification on the HAM10000 dataset

the dropout technique [19] to prevent the model from overfitting with a dropout value of 0.50 for the hidden layers. We used categorical cross-entropy [20] as the loss function for the network. We used Adam optimizer [21] for minimizing the loss function. The values of these hyperparameters were selected by using the grid search approach. From Figure 5.7, we can observe that after a certain number of epochs, the validation and training loss becomes constant after decreasing from the starting point, which is when we stopped the training process.

5.4.2 Classification Result

Ham10000 is a highly imbalanced dataset. For this, we used four different well-known evaluation metrics based on the confusion matrix for the proper evaluation of our proposed approach. They are accuracy, precision, recall, and F1-score [20]. Together, they evaluate the performance of any classification model on a class-imbalanced dataset. All the metrics have a positive correlation with the efficacy of the classification model. The higher the score, the better the model.

From Table 5.1, we can see that we have achieved not only high accuracy, but we have also achieved high precision, recall, and f1-score. This suggests that our proposed approach has achieved satisfactory performance in classifying different kinds of skin cancers from various complex skin lesions. The reason behind such classification accuracy mainly results from resolving the class imbalance using Smote and efficient and prominent feature extraction from the complex structure of skin lesions by the modified EfficientNetB0 architecture specifically tailored for skin cancer classification.

Table 5.1 Classification performance of our proposed approach using different evaluation metrics

Class Name	Accuracy (%)	Precision (%)	Recall (%)	F-Score (%)
Actinic Keratoses	99.20	98.00	99.00	98.49
Basal Cell-Carcinoma	97.86	93.00	94.00	93.49
Benign Keratosis-Like Lesion	92.89	94.00	93.00	93.49
Dermato-Fibroma	99.70	99.00	99.98	99.48
Melanocytic-Nevi	95.47	94.00	95.00	94.49
Melanoma	90.90	95.00	91.00	92.95
Vascular Lesions	99.90	98.00	97.98	97.98

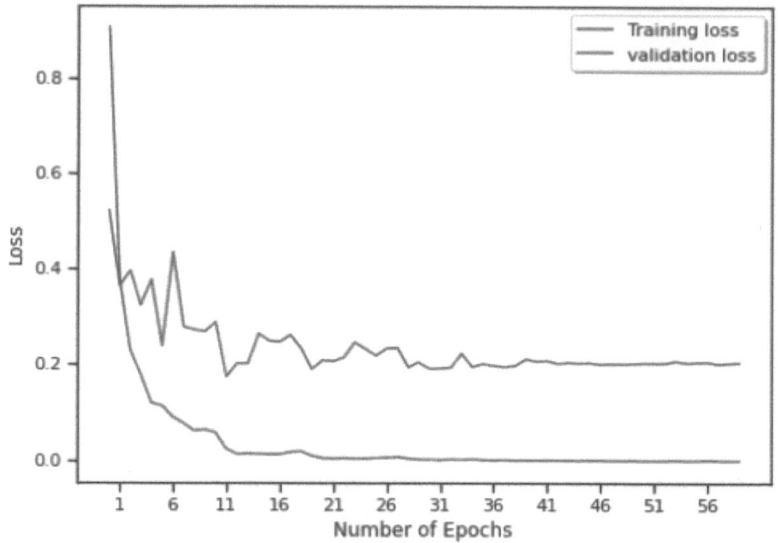

Figure 5.6 Graphical depiction of training and validation loss for our architecture

From Table 5.2, we can observe that our proposed approach has outperformed all other state-of-the-art approaches such as MobileNet, InceptionV3, ResNET50, Ensemble ResNet50, and Inception-ResNet for performing skin classification in the HAM10000 dataset.

Smote has performed significantly in creating the synthetic samples for making the training process unbiased and the modified EfficientNetB0 architecture extracted complex and intricate but useful features from various skin lesions in a remarkable way. Finally, a well-tuned MLP has been able to perform skin cancer classification based on the extracted feature in a more accurate and precise way. Furthermore, in the future, we will try to improve our classification accuracy a little bit more by introducing different kinds of segmentation approaches.

Table 5.2 Comparison with other state-of-the-art approaches for skin cancer classification on the HAM10000 dataset based on different evaluation metrics

Name of the approach	Accuracy (%)	Precision (%)	Recall (%)	F-Score (%)
InceptionV3 [22]	89.81	88.00	75.00	81.00
Inception-ResNet [25]	83.96	72.00	69.29	69.86
ResNet 50 [23]	91.00	83.00	82.00	83.00
Ensemble ResNet 50 [24]	93.00	86.00	88.00	84.00
MobileNet [8]	92.70	87.00	81.00	84.00
SMOTE-EfficientNetB0	**96.56**	**95.85**	**95.70**	**95.77**

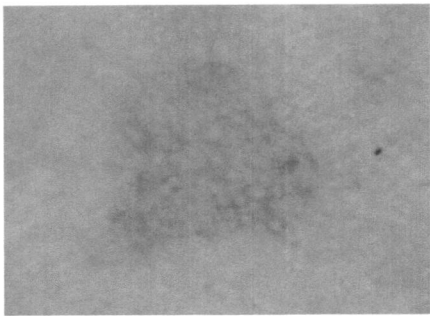 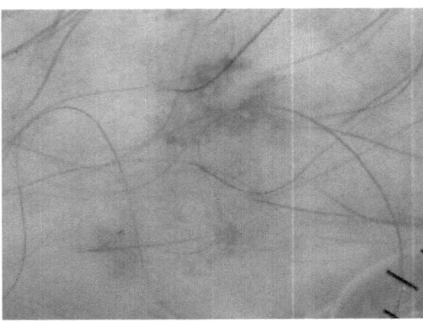

(a) Misclassification due to Skin Color   (b) Misclassification due to Hair

Figure 5.7 Understanding the misclassifications

5.4.3 Understanding the Misclassifications

After investigating the misclassifications, it turned out that most happened for two reasons. The first one is the inability to differentiate the skin part and the lesion part. This is illustrated in Figure 5.7(a). The second cause of the misclassifications is hair. Hairy images were more likely to be misclassified when our scheme was applied. One such sample is illustrated in Figure 5.7(b).

5.5 CONCLUSION

In this chapter, we presented a novel approach consisting of SMOTE and a modified EfficientNet for the diagnosis and classification of skin cancers in a more accurate and precise manner. Smote has been used to resolve the class imbalance in the training phase and modified EfficientNet has been used for the extraction of better quality skin lesion features from a heterogeneous collection of skin lesion images. Our method has achieved an accuracy of 96.7%, which surpasses the accuracy of all other benchmark methods. Not only has our proposed approach achieved better accuracy, but it has also secured better precision, recall, and f1-score, which points to the remarkable performance achieved by our proposed approach. Some skin cancers will be more frequent than others. For this reason, there will always exist a class imbalance in the cancer dataset. In the future, our proposed approach can persuade researchers to use a similar approach for the detection of other types of cancer in an imbalanced dataset. The limitation of this work is that SMOTE was necessary for better recognition in this scheme. If more significant features can be extracted with proper preprocessing steps, such as segmentation, hair removal, and skin toning, better performance is achievable without SMOTE or any augmentation.

Bibliography

[1] Diepgen, T. & Mahler, V. The epidemiology of skin cancer. *British Journal of Dermatology.* **146** pp. 1-6 (2002)

[2] WHO World Health Organization Cancer Fact Sheet. (2009)

[3] Whited, J. & Grichnik, J. Does this patient have a mole or a melanoma? *JAMA.* **279**, 696-701 (1998)

[4] Vestergaard, M., Macaskill, P., Holt, P. & Menzies, S. Dermoscopy compared with naked eye examination for the diagnosis of primary melanoma: A meta-analysis of studies performed in a clinical setting. *British Journal of Dermatology.* **159**, 669-676 (2008)

[5] Johr, R. Dermoscopy: Alternative melanocytic algorithms—the ABCD rule of dermatoscopy, Menzies scoring method, and 7-point checklist. *Clinics in Dermatology.* **20**, 240-247 (2002)

[6] Goceri, E. Skin disease diagnosis from photographs using deep learning. *ECCO-MAS Thematic Conference on Computational Vision and Medical Image Processing.* pp. 239-246 (2019)

[7] Esteva, A., Kuprel, B., Novoa, R., Ko, J., Swetter, S., Blau, H. & Thrun, S. Dermatologist-level classification of skin cancer with deep neural networks. *Nature.* **542**, 115-118 (2017)

[8] Mohamed, E. & El-Behaidy, W. Enhanced skin lesions classification using deep convolutional networks. *2019 Ninth International Conference on Intelligent Computing and Information Systems (ICICIS).* pp. 180-188 (2019)

[9] Al-Masni, M., Kim, D. & Kim, T. Multiple skin lesions diagnostics via integrated deep convolutional networks for segmentation and classification. *Computer Methods and Programs in Biomedicine.* **190** pp. 105351 (2020)

[10] Tan, M. & Le, Q. EfficientNet: Rethinking model scaling for convolutional neural networks. *International Conference on Machine Learning.* pp. 6105-6114 (2019)

[11] Japkowicz, N. & Stephen, S. The class imbalance problem: A systematic study. *Intelligent Data Analysis.* **6**, 429-449 (2002)

[12] Chawla, N., Bowyer, K., Hall, L. & Kegelmeyer, W. SMOTE: synthetic minority over-sampling technique. *Journal of Artificial Intelligence Research.* **16** pp. 321-357 (2002)

[13] Tschandl, P., Rosendahl, C. & Kittler, H. The HAM10000 dataset, a large collection of multi-source dermatoscopic images of common pigmented skin lesions. *Scientific Data.* **5**, 1-9 (2018)

[14] Fernández, A., Garcia, S., Herrera, F. & Chawla, N. SMOTE for learning from imbalanced data: Progress and challenges, marking the 15-year anniversary. *Journal of Artificial Intelligence Research.* **61** pp. 863-905 (2018)

[15] Tan, M., Chen, B., Pang, R., Vasudevan, V., Sandler, M., Howard, A. & Le, Q. Mnasnet: Platform-aware neural architecture search for mobile. *Proceedings of the IEEE/CVF Conference on Computer Vision and Pattern Recognition.* pp. 2820-2828 (2019)

[16] Popescu, M., Balas, V., Perescu-Popescu, L. & Mastorakis, N. Multilayer perceptron and neural networks. *WSEAS Transactions on Circuits and Systems.* **8**, 579-588 (2009)

[17] Bouchard, G. Efficient bounds for the softmax function and applications to approximate inference in hybrid models. *Proceedings of the Presentation at the Workshop for Approximate Bayesian Inference in Continuous/Hybrid Systems at Neural Information Processing Systems (NIPS), Meylan, France.* **31** (2008)

[18] GitHub. Keras: Deep learning for humans. `https://github.com/keras-team/keras`, Accessed: 2021-07-15

[19] Srivastava, N., Hinton, G., Krizhevsky, A., Sutskever, I. & Salakhutdinov, R. Dropout: A simple way to prevent neural networks from overfitting. *The Journal of Machine Learning Research.* **15**, 1929-1958 (2014)

[20] Zhang, Z. & Sabuncu, M. Generalized cross entropy loss for training deep neural networks with noisy labels. *32nd Conference on Neural Information Processing Systems (NeurIPS).* (2018)

[21] Kingma, D. & Ba, J. Adam: A method for stochastic optimization. *ArXiv Preprint ArXiv:1412.6980.* (2014)

[22] Goutte, C. & Gaussier, E. A probabilistic interpretation of precision, recall and F-score, with implication for evaluation. *European Conference on Information Retrieval.* pp. 345-359 (2005)

[23] Deif, M. & Hammam, R. Skin lesions classification based on deep learning approach. *Journal of Clinical Engineering.* **45**, 155-161 (2020)

[24] Le, D., Le, H., Ngo, L. & Ngo, H. Transfer learning with class-weighted and focal loss function for automatic skin cancer classification. *ArXiv Preprint ArXiv:2009.05977.* (2020)

[25] Mureşan, H. Skin Lesion Diagnosis Using Deep Learning. *2019 IEEE 15th International Conference on Intelligent Computer Communication and Processing (ICCP)*. pp. 499-506 (2019)

CHAPTER 6

Advanced GradCAM++: Improved Visual Explanations of CNN Decisions in Diabetic Retinopathy

Md. Shafayat Jamil

Computational Color and Spectral Image Analysis Lab, Computer Science and Engineering Discipline, Khulna University, Khulna, Bangladesh

Sirdarta Prashad Banik

Computational Color and Spectral Image Analysis Lab, Computer Science and Engineering Discipline, Khulna University, Khulna, Bangladesh

G. M. Atiqur Rahaman

Computational Machine Perception and Interaction Lab, Center for Applied Research on Autonomous Censor Systems, Örebro University, Sweden

Sajib Saha

Australian e-Health Research Centre, CSIRO, Floreat, Australia

CONTENTS

ONVOLUTIONAL NEURAL NETWORK (CNN)-based methods have achieved state-of-the-art performance in solving several complex computer vision problems including assessment of diabetic retinopathy (DR). Despite this, CNN-based methods are often criticized as "black box" methods for providing limited to no understanding about their internal functioning. In recent years there has been increased interest in developing explainable deep learning models, and this chapter is an effort in that direction in the context of DR. Based on one of the best performing methods, called Grad-CAM++, we propose Advanced Grad-CAM++ to provide further improvement in visual explanations of CNN model predictions (when compared to Grad-CAM++), in terms of better localization of DR pathology as well as explaining occurrences of multiple DR pathology types in a fundus image. By keeping all the layers and operations as is, the proposed method adds an additional non-learnable bilateral convolutional layer between the input image and the very first learnable convolutional layer of Grad-CAM++. Experiments were conducted on fundus images collected from publicly available sources, namely EyePACS and DIARETDB1. The intersection over Union (IoU) score between the ground truth and heatmap produced by the methods were used to quantitatively compare performance. The overall IoU score for Advanced Grad-CAM++ is 0.179, whereas for Grad-CAM++ it is 0.161. Thus, an 11.18% improvement in agreement with the ground truths by the proposed method is inferable.

6.1 INTRODUCTION

Diabetic retinopathy (DR), which affects the retina, is one of the most common complications of uncontrolled diabetes [23]. DR is a major cause of irreversible visual loss in the world [1]. According to [2], more than 415 million people or 1 in 11 adults worldwide are Diabetes affected. Around 40-45% of patients with Diabetes have a chance of getting affected with DR at some point in their lives. Early detection is the key to reducing the risk of blindness. It is recommended that patients with Diabetes need to be evaluated for DR twice a year. The present clinical practice requires expert ophthalmologists to review color images of the fundus to detect DR. It is becoming increasingly difficult to provide comprehensive eye care to an ever-growing population. It is not only prohibitively expensive, but it is also infeasible in many socio-economic situations as well. Over the past few decades, there has been a growing interest in developing machine learning-based methods for automated diagnosis of fundus images in order to address these issues and overcome them.

In the last several years, there has been a tremendous increase in interest in deep learning as an advanced machine learning approach. Convolutional Neural Networks (CNNs), a type of deep learning approach, have achieved tremendous success in a variety of computer vision and image classification applications, including fast, reliable computer-assisted diagnosis of DR. RESNET [4], Inception [5], VGG-16 [6] etc. are examples of CNN models that are trained over a large dataset and used for analysis of images in vast application areas. These models are trained using different feature spaces of the training dataset for object detection, pattern recognition, object visualization, speech recognition, image restoration, [7] etc. CNN-based systems are expected to be faster and more accurate and robust with the availability of larger datasets and powerful GPU implementations [3].

Whilst CNN-based models have demonstrated impressive classification performance, the interpretability of the CNN's decision-making process still remains insufficient. A number of visualization techniques, including Saliency, SmoothGrad, GradCAM, GradCAM++, and ScoreCAM, have been presented recently to assist in understanding the internal functioning of CNNs. Despite the fact that these strategies have been found to be effective to a certain level, there is still room for development [8].

Among the list of visualization techniques, GradCAM++ is one of the best performing methods. In this chapter, we propose an advanced version of GradCAM++ in the context of DR, which ensures better explainability as well as higher agreement with experts' annotations. By keeping all the layers and operations of GradCAM++ as is, the proposed method adds an additional not-learnable bilateral convolutional layer between the input image and the very first learnable convolutional layer of Grad-CAM++.

6.2 BACKGROUND

6.2.1 Convolutional Neural Networks (CNNs)

The convolutional neural network (CNN) [20] is usually seen as interconnected layers of neurons, which share information. Raw data (e.g., an image) is fed into the network and the data is then generated by each layer in succession (for example, the first layer may represent the location and orientation of edges within an image, while successive layers may deal with higher levels of abstraction). Finally, neuron output is activated and data are classified. The deep CNN architectures are typically made of numerous layers of convolution. There are numerous convolutional filters applied on the image at each convolutional layer. Each convolutional layer at level L takes an image of dimension dL and applies N number of filters to produce N number of feature maps. This convolution operation produces an N-dimensional image, one dimension per filter. This N-dimensional image is then taken as input to the next convolutional layer at level $d(L+1)$. This process continues for several convolutional layers. Finally, several fully connected neural network layers are added on top of the convolutional layers. Typically, the final layer consists of a soft-max classifier.

Among different CNN model architectures, VGG-16 is one of the simplest and most frequently used models [22]. In VGG-16, there are 16 layers with learnable weights that include 13 convolutional layers, 3 fully connected layers and pooling layers. In the last layer, they are connected by fully connected layers and classified by a softmax unit in classes that do not overlap. The model takes $224 \times 224 \times 3$ sized input RGB color images. All hidden layers use rectified non-linearity functions (ReLU).

6.2.2 Visualizing CNNs

Visualization techniques are used to realize the outputs that we get from CNN models [21]. Visualization techniques can generally be classified into two groups:

gradient-orientated and class-oriented activation maps (CAM). Some of the commonly used gradient based approaches include saliency and SmoothGrad. A saliency map is a gradient based method which works with the neuron in the final fully connected classification layer and has to be maximized to visualize the class of interest [10]. It also uses a single back-propagation pass through any classification of ConvNet and this method can be used for weakly supervised localization [10]. SmoothGrad is another gradient based visualization technique that is slightly better than the saliency map method. To reduce noise, this method adds noise to the input images, takes the average result of sensitivity maps for each input image, and uses the common regularization technique of adding noise at training time for "denoising" effect [8]. Some of the widely known CAM based approaches include CAM, grad-CAM, Grad CAM++, Score CAM. The main methodology of Grad CAM, Grad CAM ++, and Score-CAM are almost similar. Grad CAM stands for Gradient-weighted Class Activation Mapping, a technique for class-discriminating image localisation that generates pixel-space gradients of images [8]. Score CAM is not gradient-dependent. It calculates the weight of each activation map based on its forward passing score on the target class; the final result is obtained by a simple combination of weights and activation maps [13]. Grad-CAM++ is a feature upgrade of Grad-CAM.

Grad CAM++ is the combination of grad Cam and Guided Backpropagation, which is generated by point-wise multiplication. Guided Backpropagation and Deconvolution do not have that much ability to show the importance of fine-grained, pixel-space gradient visualization techniques [9]. In [9], the authors modified the base model to remove all fully connected layers at the end and they included a tensor product which was followed by the softmax function. The Global average pooled layer was taken as input and the output was a probability prediction for each class. Finally, point-wise multiplication was operated to generate the heat map.

6.3 PROPOSED VISUALIZATION TECHNIQUE

The proposed method has been inspired by GradCAM++ [9], one of the best performing CNN visualization techniques. In order to further improve the performance of GradCAM++, in this work we propose adding a bilateral convolutional layer prior to the trainable convolutional layers. GradCAM++ largely relies upon gradient information, and gradient computations are widely known to be prone to noise. The primary idea of using the bilateral convolutional layer is to minimize the effect of noise. A bilateral convolutional layer inside a CNN has already been used by others in [14] and [15]. However, in contrast to [14] and [15], who applied them in the intermediate outputs produced by a CNN, in this work we apply it directly on the input image. An overview of our proposed visualization technique is shown in Fig. 10.1

There are many available linear and non-linear filters that could be used to minimize the effects of noise. Some of the widely known filters include the median filter, Gaussian filter, bilateral filter, average filter, etc. The bilateral filter is one of the best performing smoothing filters that preserves the edges. In bilateral filtering, for each pixel of an image, an approximate intensity value is computed from the nearby pixels. Further the intensity value of the pixel is replaced by the computed approximate

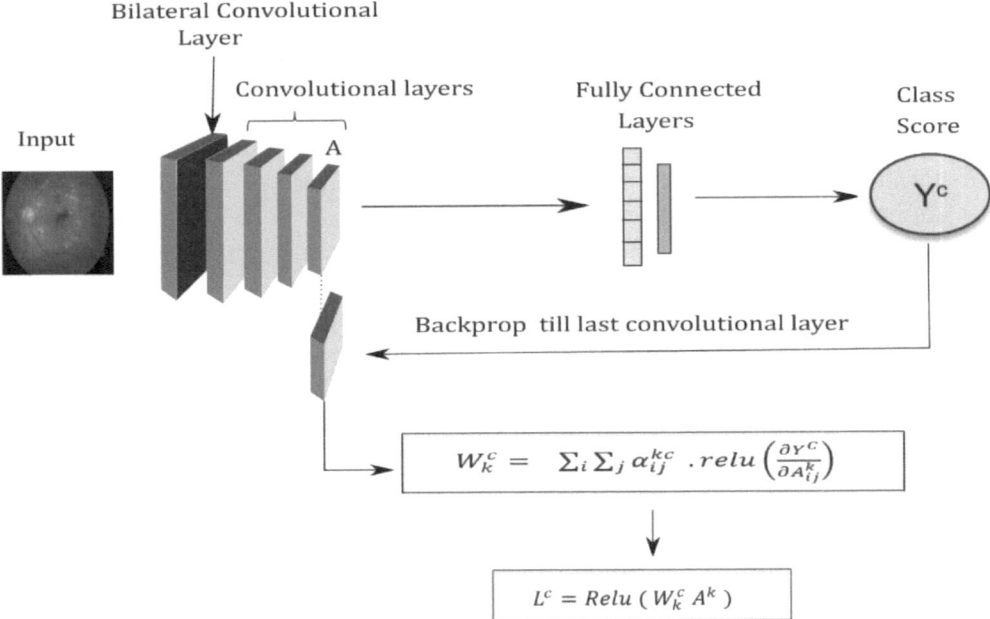

Figure 6.1 An overview of the proposed visualization technique.

intensity value [16]. Bilateral filtering can be mathematically explained as below:

$$I^{filtered}(X) = \frac{1}{W^p} \sum_{x_i \in \Omega} I(x_i) f_r(||I(x_i) - I(x)||) g_s(||x_i - x||) \qquad (6.1)$$

Here I is an input image, $I^{filtered}$ is the output image after filtering, x are the coordinate of a current pixel, f_r is the range kernel, and g_s is the domain kernel. -Here, the normalization factor is defined by:

$$W_p = \sum_{x_i \in \Omega} f_r(||I(x_i) - I(x)||) g_s(||x_i - x||) \qquad (6.2)$$

The weights used here are based on the Gaussian distribution [11]. According to [15], the formulation of a bilateral filter is simple and replacing the intensity value of a pixel with the weighted average of nearby pixels helps to achieve information about behavior. It also helps to adapt the application-specific requirements.

To generate the explanation map, we follow the Grad-CAM ++ formula. It can improve the predicted category's accuracy over Grad-CAM, hence enhancing the model's reliability [9]. The behavior of the confidence value (of the deep network) for that particular class, in which both the original and integrated images of the heatmap are displayed, may be analyzed to identify which method generates the specific heatmap for a given category. The gradient of the class score (Yc) is computed in this method in relation to the feature map of the final convolution layer, that is:

$$\frac{\partial y^c}{\partial A_{ij}^k} = 1 \qquad if A_{ij}^k = 1 \quad and \quad \frac{\partial y^c}{\partial A_{ij}^k} = 0 \qquad if A_{ij}^k = 0. \tag{6.3}$$

Further, a weighted average of the pixel-wise gradients W_k^c is computed where the importance of a particular activation map A^k is captured. It is defined by:

$$W_k^c = \sum_i \sum_j \alpha_{ij}^{kc} relu\left(\frac{\partial y^c}{A_{ij}^k}\right). \tag{6.4}$$

Taking a further partial derivative $A_{i}j^k$ and rearranging terms in [9]:

$$\alpha_{ij}^{kc} = \frac{\frac{\partial^2 y^c}{(\partial A_{ij}^k)^2}}{2\frac{\partial^2 y^c}{(\partial A_{ij}^k)} + \sum_a \sum_a A_{ij}^k \{\frac{\partial^3 y^c}{(\partial A_{ij}^k)^3}\}}. \tag{6.5}$$

Substituting equation (5) in equation (4), the authors get the following Grad-CAM++ weights in [9]:

$$W_k^c = \sum_i \sum_j [\frac{\frac{\partial^2 y^c}{(\partial A_{ij}^k)^2}}{2\frac{\partial^2 y^c}{(\partial A_{ij}^k)} + \sum_a \sum_a A_{ij}^k \{\frac{\partial^3 y^c}{(\partial A_{ij}^k)^3}\}}].relu\left(\frac{\partial y^c}{A_{ij}^k}\right). \tag{6.6}$$

Finally, the class discriminative saliency map is generated following the equation:

$$L^c = Relu(W_k^c A^k). \tag{6.7}$$

6.4 EXPERIMENTS AND RESULTS

This section discusses the performance of the visualization method in detail, as well as the CNN model and experimental results.

6.4.1 Training CNN model for disease level grading of DR

A VGG-16 model was trained on the EyePacs [17] dataset to perform 'mild non-proliferative diabetic retinopathy (NPDR)', 'moderate NPDR', 'severe NPDR' and 'proliferative DR (PDR)' disease-level grading. EyePacs had a total of 35,126 annotated images which were graded into 'No DR', 'mild NPDR', 'moderate NPDR', 'severe NPDR' and 'PDR'.

The data were unbalanced and most of them were 'No DR' category. The 'PDR' and 'severe NPDR' categories had the fewest images, respectively 708 and 873. In order to ensure proper training and testing of the CNN, we aimed to use almost the same number of images for each category. Thus, a set of 3360 randomly chosen images were used for this experiment. Eighty percent of these images were used for training and the remaining 20% were used for testing. We obtained an accuracy of 91.38%. Figure 6.2(a) shows the training and test accuracy as well as the loss curve. Figure 6.2(b) shows the ROC curve of the four DR categories.

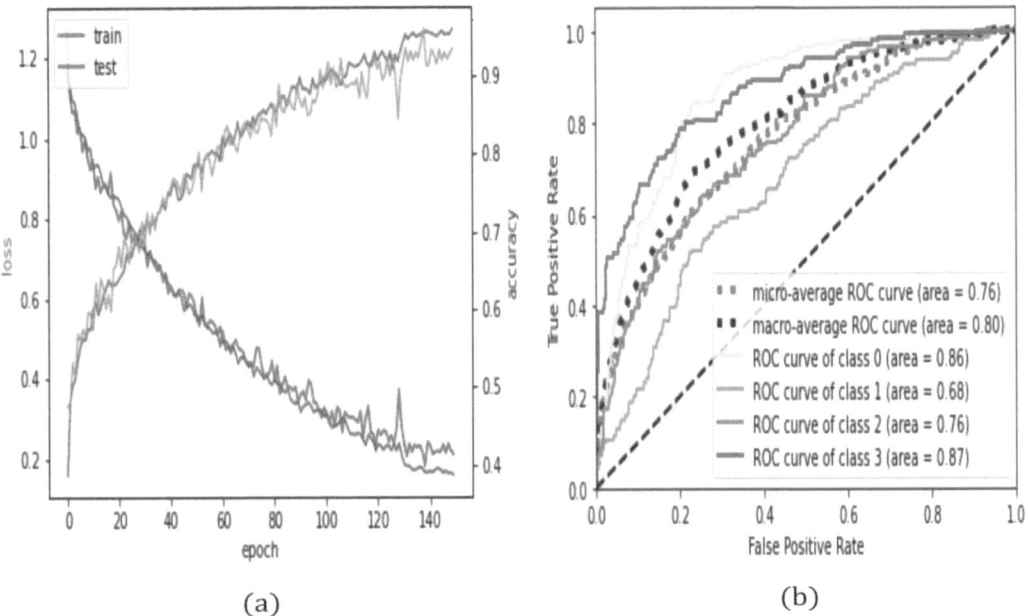

Figure 6.2 (a) Training accuracy, validation accuracy and loss of VGG16 in Diabetic Retinopathy detection. (b) ROC curves showing the accuracy for four different DR classes.

6.4.2 Visualizing CNN through GradCAM++ and proposed method

In this section we visualize the VGG-16 decision using both GradCAM++ and the proposed method, and compare the visual maps produced by the two methods. These visual maps highlight the input image regions which are discriminant and contribute to the CNN decision-making process. Fig. 6.3 shows the examples of side-by-side comparison of the visual maps produced by GradCAM++ and the proposed method on sample images from DIARETDB1.

In Fig. 6.3, the fundus image represents the main input image of the DIARETDB1 database and Ground truth inspects the regions of hard exudates, soft exudates, microaneurysms and highlights these regions in the image. Our investigation of these two visual maps shows that this proposed method inspects features better than the GradCAM++ visualization method. Those regions that are missed in inspection by the GradCAM++ method, are slightly better detected by Advanced GradCAM++ when comparing the images of these two methods with the ground truth images that we see in Fig. 6.3.

For quantitative evaluation, we computed the Intersection over union (IOU) [19] of three mutually exclusive categories between the ground truth and explanation maps (GradCAM++ and Advanced GradCAM++). IOU is a numerical score which is computed from 0 to 1 and specifies the amount of overlap between the ground truth and explanation maps. IOU is computed for both the GradCAM++ and Advanced GradCAM++ methods. Finally, we computed the mean IOU score over 3 Categories

Fundas Image   Ground Truth   GradCam++   Proposed Technique

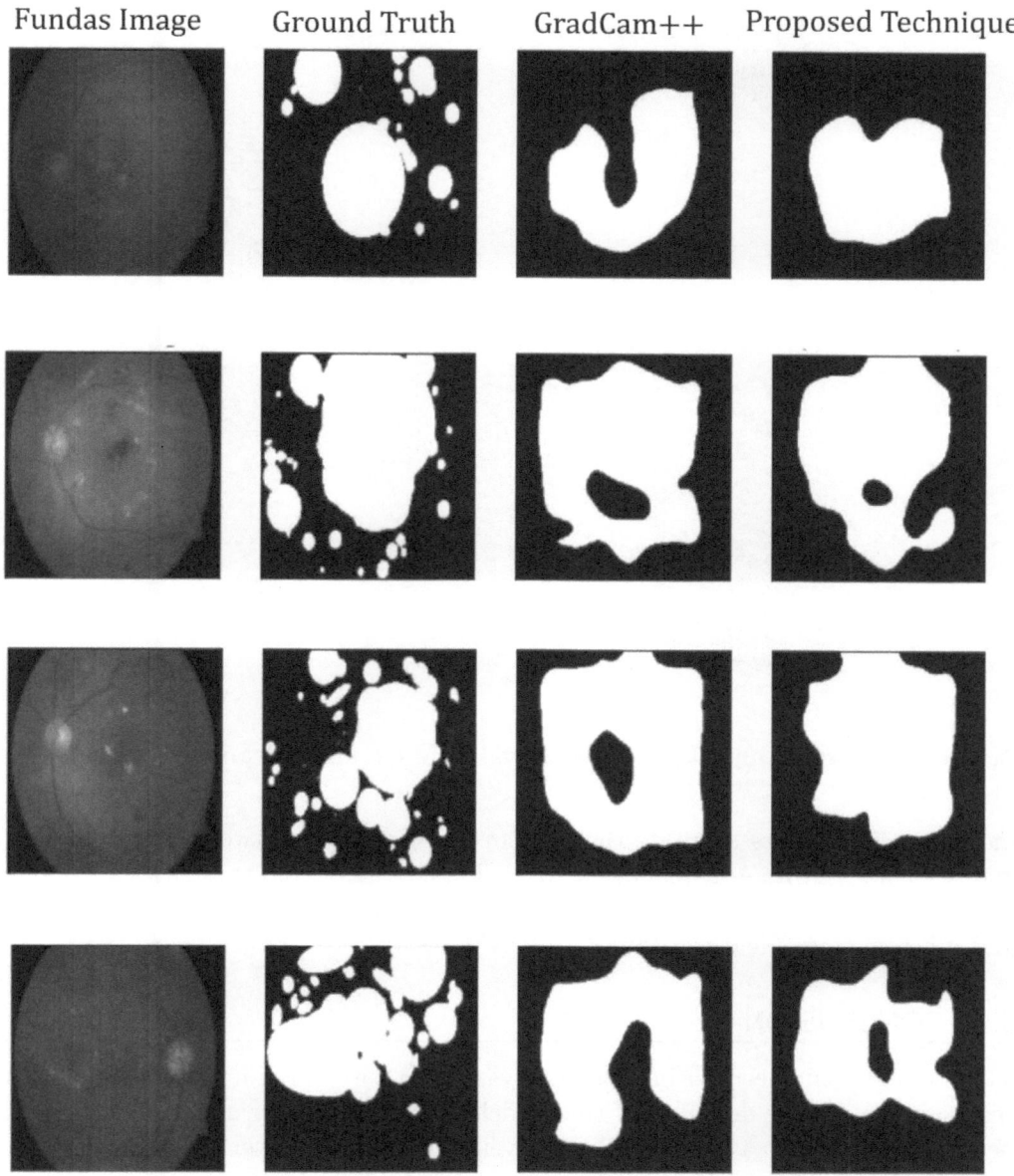

Figure 6.3 Qualitative evaluation of GradCAM++ and Advanced GradCAM++.

(89 images, 80 images and 40 images). Fig. 6.4 shows a quantitative comparison between the computed mean IOU of both methods over 3 categories.

Inspecting the quantitative evidence, we observed the following for all three categories: the mean IOU score for Advanced GradCAM++ is slightly better than the IOU score for GradCAM++, such as for the 40-image category, the IOU of Grad-CAM++ is 0.245 and the IOU of Advanced GradCAM++ is 0.273. This means the overlapped area between ground truth and the Advanced GradCAM++ explanation

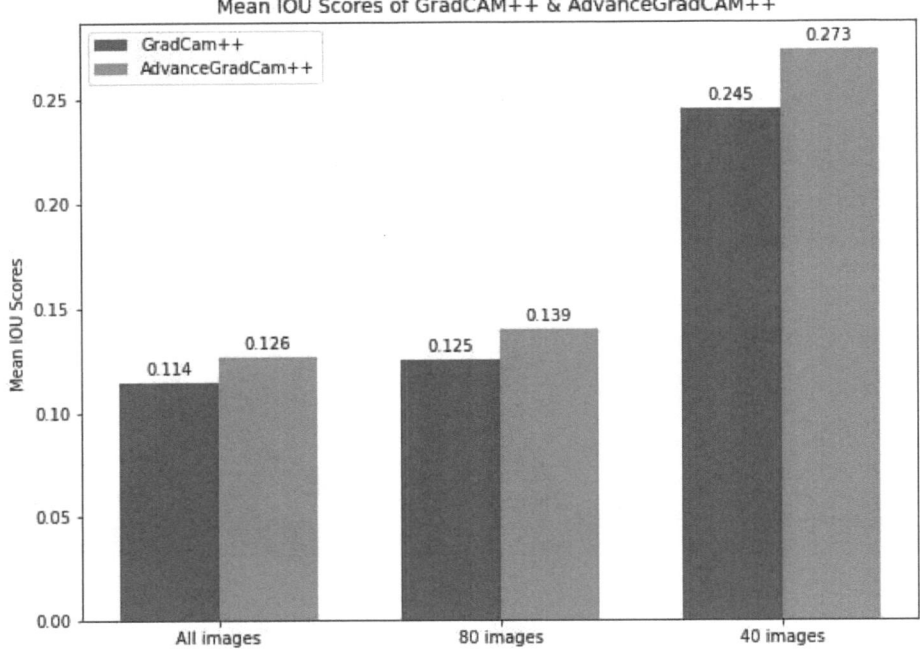

Figure 6.4 Plot of mean IoU scores for GradCAM++ and Advanced GradCAM++.

map is 10.256% more accurate than the overlapping area between the ground truth and the GradCAM++ explanation map. These results confirm that the Advanced GradCAM++ method shows qualitatively and quantitatively better discrimination and inherent image features than GradCAM++.

6.5 CONCLUSION

There is a growing need for interpretive systems in automated medical image analysis that can explain the decisions of the models. Existing visualization methods need to be improved to reach a high confidence level in health care systems. In this chapter, an extended version of the GradCAM++ model is presented in the context of Diabetic Retinopathy detection. The results suggest that the performance of the gradient-based model GradCAM++ can be improved by adding a bilateral convolutional layer. It helps to provide a better visual explanation map to increase our understanding of CNN decisions. As a result, the professional trust and integration of the models into clinical settings such as phase support, diagnosis and treatment can be raised. A further area for exploration is to extend our proposed technique to handle different multiple classes and network architectures.

Bibliography

[1] Wan, S., Liang, Y., & Zhang, Y. (2018). Deep Convolutional Neural Networks for Diabetic Retinopathy Detection by Image Classification. *Computers & Electrical Engineering* , 72, 274–282. http://doi.org/10.1016/j.compeleceng.2018.07.042

[2] Gargeya, R., & Leng, T. (2017). Automated Identification of Diabetic Retinopathy Using Deep Learning. *Ophthalmology* , 124(7), 962–969. http://doi.org/10.1016/j.ophtha.2017.02.008

[3] Zeiler, M. D., & Fergus, R. (2014). Visualizing and Understanding Convolutional Networks. *Computer Vision – ECCV 2014*, 818–833. http://doi.org/10.1007/978-3-319-10590-1_53

[4] Wu, Z., Shen, C., & van den Hengel, A. (2019). Wider or Deeper: Revisiting the ResNet Model for Visual Recognition. *Pattern Recognition, 90*, 119–133. http://doi.org/10.1016/j.patcog.2019.01.006

[5] Szegedy, C., Vanhoucke, V., Ioffe, S., Shlens, J., & Wojna, Z. (2016). Rethinking the Inception Architecture for Computer Vision. *2016 IEEE Conference on Computer Vision and Pattern Recognition (CVPR)*. http://doi.org/10.1109/cvpr.2016.308

[6] Simonyan, K., & Zisserman, A. (2015, April 10). Very Deep Convolutional Networks for Large-Scale Image Recognition. Retrieved July 8, 2021, from https://export.arxiv.org/abs/1409.1556

[7] Tammina, S. (2019). Transfer Learning Using VGG-16 with Deep Convolutional Neural Network for Classifying Images. *International Journal of Scientific and Research Publications (IJSRP), 9*(10). http://doi.org/10.29322/ijsrp.9.10.2019.p9420

[8] Smilkov, D., Thorat, N., Kim, B., Viégas, F., & Wattenberg, M. (2017, June 12). SmoothGrad: Removing Noise by Adding Noise. Retrieved July 8, 2021, from https://arxiv.org/abs/1706.03825

[9] Chattopadhay, A., Sarkar, A., Howlader, P., & Balasubramanian, V. N. (2018). Grad-CAM++: Generalized Gradient-Based Visual Explanations for Deep Convolutional Networks. *2018 IEEE Winter Conference on Applications of Computer Vision (WACV)*. http://doi.org/10.1109/wacv.2018.00097

[10] Simonyan, K., Vedaldi, A., & Zisserman, A. (2014, April 19). Deep Inside Convolutional Networks: Visualising Image Classification Models and Saliency Maps. Retrieved July 8, 2021, from https://arxiv.org/abs/1312.6034v2

[11] Omeiza, D., Speakman, S., Cintas, C., & Weldermariam, K. (2019, August 3). Smooth Grad-CAM++: An Enhanced Inference Level Visualization Technique for Deep Convolutional Neural Network Models. Retrieved July 8, 2021, from https://arxiv.org/abs/1908.01224v1

[12] Selvaraju, R. R., Cogswell, M., Das, A., Vedantam, R., Parikh, D., & Batra, D. (2017). Grad-CAM: Visual Explanations from Deep Networks via Gradient-Based Localization. *2017 IEEE International Conference on Computer Vision (ICCV)*. http://doi.org/10.1109/iccv.2017.74

[13] Wang, H., Wang, Z., Du, M., Yang, F., Zhang, Z., Ding, S., ... Hu, X. (2020). Score-CAM: Score-Weighted Visual Explanations for Convolutional Neural Networks. *2020 IEEE/CVF Conference on Computer Vision and Pattern Recognition Workshops (CVPRW)*. http://doi.org/10.1109/cvprw50498.2020.00020

[14] Gadde, R., Jampani, V., Kiefel, M., Kappler, D., & Gehler, P. V. (2016). Superpixel Convolutional Networks Using Bilateral Inceptions. *Computer Vision – ECCV 2016*, 597–613. http://doi.org/10.1007/978-3-319-46448-0_36

[15] Jampani, V., Kiefel, M., & Gehler, P. V. (2016). Learning Sparse High Dimensional Filters: Image Filtering, Dense CRFs and Bilateral Neural Networks. *2016 IEEE Conference on Computer Vision and Pattern Recognition (CVPR)*. http://doi.org/10.1109/cvpr.2016.482

[16] Paris, S., Kornprobst, P., Tumblin, J. T., & Durand, F. (2008). Bilateral Filtering: Theory and Applications. *Foundations and Trends® in Computer Graphics and Vision, 4*(1), 1–75. http://doi.org/10.1561/0600000020

[17] Cuadros, J., & Bresnick, G. (2009). EyePACS: An Adaptable Telemedicine System for Diabetic Retinopathy Screening. *Journal of Diabetes Science and Technology, 3*(3), 509–516. http://doi.org/10.1177/193229680900300315

[18] Kauppi, T., Kalesnykiene, V., Kamarainen, J.-K., Lensu, L., Sorri, I., Raninen, A., ... Pietila, J. (2007). The DIARETDB1 Diabetic Retinopathy Database and Evaluation Protocol. *Procdings of the British Machine Vision Conference 2007*. http://doi.org/10.5244/c.21.15

[19] Hofesmann, E. (2021, March 1). IoU a Better Detection Evaluation Metric. Retrieved July 8, 2021, from https://towardsdatascience.com/iou-a-better-detection-evaluation-metric-45a511185be1

[20] LeCun, Y., Bengio, Y., & Hinton, G. (2015). Deep Learning. *Nature, 521*(7553), 436–444. http://doi.org/10.1038/nature14539

[21] Vaghjiani, D., Saha, S., Connan, Y., Frost, S., & Kanagasingam, Y. (2020). Visualizing and Understanding Inherent Image Features in CNN-based Glaucoma Detection. *2020 Digital Image Computing: Techniques and Applications (DICTA)*. http://doi.org/10.1109/dicta51227.2020.9363369

[22] Saha SK, Fernando B, Xiao D, Tay-Kearney ML, Kanagasingam Y.(2016). Deep Learning for Automatic Detection and Classification of Microaneurysms, Hard and Soft Exudates, and Hemorrhages for Diabetic Retinopathy diagnosis. Investigative Ophthalmology & Visual Science.;57(12):5962-.

[23] Saha, S. K., Fernando, B., Cuadros, J., Xiao, D., & Kanagasingam, Y. (2018). Automated Quality Assessment of Colour Fundus Images for Diabetic Retinopathy Screening in Telemedicine. *Journal of Digital Imaging, 31*(6), 869–878. http://doi.org/10.1007/s10278-018-0084-9

CHAPTER 7

Bangla Sign Language Recognition Using a Concatenated BdSL Network

Thasin Abedin

Islamic University of Technology (IUT), Gazipur, Bangladesh

Khondokar S. S. Prottoy

Islamic University of Technology (IUT), Gazipur, Bangladesh

Ayana Moshruba

Islamic University of Technology (IUT), Gazipur, Bangladesh

Safayat Bin Hakim

Islamic University of Technology (IUT), Gazipur, Bangladesh

CONTENTS

SIGN LANGUAGE is the only medium of communication for the physically impaired community. Communication with the general population is thus always a challenge for this minority group. Especially in Bangla sign language (BdSL), there are 38 alphabets with some having nearly identical symbols. As a result, in BdSL recognition, the posture of the hand is an important factor in addition to visual features extracted from a traditional Convolutional Neural Network (CNN). In this chapter, a novel architecture, "Concatenated BdSL Network," is proposed which consists of a CNN based image network and a pose estimation network. While the image network gets the visual features, the relative positions of hand keypoints are taken by the pose estimation network to obtain the additional features to deal with the complexity of the BdSL symbols. A score of 91.51% was achieved by this novel approach in the test set and the effectiveness of the additional pose estimation network is suggested by the experimental results.

7.1 INTRODUCTION

Basically visual and manual modes are used in Sign Language(SL) in order to communicate. It is a rather complex mechanism specially for the general speaking and talking population. So with the progress in the field of computer vision, much research has been done to detect sign language. But the amount of research is rather small in the case of detection of BdSl, [1, 2, 3, 4] being worth mentioning. There are 38 symbols of which 9 are vowels and 27 are consonants in BdSL. The Machine Learning classifier, Support Vector Machine (SVM), was used by a lot of researchers to detect each alphabet from the hand gestures. But image classifiers, like CNN, are now popular among scientists to detect Sign Languages from images. The other methods are Principal Component Analysis (PCA), skeleton detection, Hidden Markov Models etc. [5, 7].

In this chapter, a novel model, a combination of an Image Network (CNN) [2, 3] and a Pose Estimation Network is proposed. A CNN model is used, which was trained for visual feature extraction from the images. A pre-trained hand key estimation model, Openpose, is used to estimate the hand key points from the image. The outputs from both the CNN and Openpose are then concatenated through 3 connected layers. The dataset used was the Bengali Sign Language Dataset obtained by the students of Bangladesh National Federation of the Deaf (BNFD).

7.2 LITERATURE REVIEW

Of the 300 sign languages existing throughout the world, not many of them have sign language recognition tools. With the advancement in Artificial Intelligence, researchers have been interested to work on sign language recognition to make life easier for the deaf-mute community. Among the other languages, most research has been performed on American Sign Language (ASL) detection. A real-time tool on the basis of desk and wearable computer based videos to recognise ASL was built by Starner et al. [10], and again, a Hidden Markov model was used in [12], which is also applicable in real time. In [11], kinetics was used for ASL recognition. Parallel Hidden Markov models were used in [13]. Sign language recognition systems have also been developing for other languages like Chinese Sign Language(CSL),Indian

Sign Language(ISL), Japanese Sign Language(JSL), Arabic Sign Language, French Sign Language(FSL), etc. In [14], a real-time model was built for ISL. A classification technique based on Eucliean distance, which is Eigen value weighted, was used by Singha et al. in [15]. For JSL recognition in [16], a hand tracking system was built based on color, and [17] showed hand feature extraction for JSL. For CSL recognition in [18], a phonemes based approach was taken. For CSL recognition, a sign component based framework was used with an accelerometer and sEMG data in [19]. These are some of the notable works in Sign Language Recognition.

Bangla Sign language recognition(BdSL) is not yet much explored by researchers. BdSL is actually quite unique. It is a modified form of American SL, British SL and Australian SL. The Support Vector Machine (SVM) [20], K-Nearest Neighbour (KNN), and Artificial Neural Network (ANN) [21] were previously the common techniques used for BdSL recognition, but in recent times, CNN is the most popular among researchers for this job. In [4], the Scale-Invariant Feature Transform (SIFT) was used to extract features and CNN for detection by SS Shanta et al. A VGG19 based CNN model for BdSL recognition was used by Abdul Muntakim Rafi and his team in [1]. In [2], a CNN model was used by Md. Sanzidul Islam et al. to recognize BdSL digits. In [3], Faster R-CNN was used by Oishee Bintey Hoque and her team for real-time BdSL recognition. A novel BdSL recognition model combining both CNN and Pose estimation is proposed in this work which has not been done by other researchers before.

7.3 METHODOLOGY

This chapter proposes a novel architecture for BdSL recognition that combines an image network with a pose estimation network.

7.3.1 Data Preprocessing

In this architecture, there are two separate inputs for the image network and the pose estimation network. For the image network input, the images are converted into grayscale. The images for the pose estimation network inputs are in BGR format. Both the image inputs are normalized by the highest level (255) and resized into 64 x 64 for the task.

7.3.2 Proposed Architecture

The proposed novel architecture consists of a separate image network and a pose estimation network. The two networks are tasked with separate but closely related purposes in terms of BdSL recognition. Our architecture builds a mechanism to efficiently use both the networks to give state-of-the-art results for Bangla sign language recognition. This chapter refers to the proposed architecture as a "Concatenated BdSL Network".

7.3.2.1 Image Network

The image network is tasked with the purpose of extracting visual features from pixel images. In this case, for the image network, CNNs were chosen for visual feature extraction as they provide better results compared to other machine learning approaches in case of feature extraction from images [1]. For most of the works on BdSL recognition, pretrained CNN models were used due training difficulties, shortage of data, etc. [4]. However, the CNN model for this architecture is trained from scratch. The efficiency of the CNN is dependent on the architectural design. To select the best architecture, there is no universal rule and it is made by choosing the right number of convolution layers and neurons.

By hypertuning various parameters, the best architecture was selected for this task. In this architecture, the CNN comprises 10 convolution layers with batch normalization for each layer, 10 "ReLu" activation layers, 4 max pooling layers, along with a single input and output layer. The convolution layers use trainable filters to detect the presence of specific features or patterns present in the input image to generate feature maps that are passed to the next layers of CNN. The "ReLu" activation layer proves to be the best choice for our image network as it converges faster than other activation functions by not activating all the neurons at the same time. It also does not saturate at the positive regions. After the activation layer, each layer is followed by a batch normalization to standardize the inputs and make the learning process stable and reduce the training time. The max pooling layers further down sample the feature maps to highlight the most prominent features to be considered only. Finally, the output of the image network is flattened to a shape of 1x1x8196.

7.3.2.2 Pose Estimation Network

The task of recognizing the Bangla sign language is very complex as there are many letters in Bangla. Moreover, there is very subtle change among the gestures of some specific symbols that are even difficult to detect with human eyes. Hence, to aid our image network for this complex task of detecting the subtle differences among different hand gestures, a pose estimation network is included in this architecture. The main purpose of this pose estimation network is to estimate the hand keypoints from images. However, to train a pose estimation model from scratch is very computationally expensive and requires a huge amount of data. For this reason, a pretrained hand keypoint estimation model from Openpose [1] is used. It estimates 21 co-ordinate points from a single hand image. So, our pose estimation network takes images as input and gives an output shape of 21x2. The final obtained output is flattened to a shape of 1x42 in a similar way to the image network output.

7.3.2.3 Concatenated BdSL Network

The features extracted from the image network and pose estimation network are complimentary to each other. As a result, combining both these features can greatly help to build a precise BdSL recognition model and tackle the complexities at the same time. To do so, the final flattened output from both the image and pose estimation networks are passed through two fully connected layers with a "Relu" activation function and concatenated together. Afterwards, the new concatenated features are passed through 3 more fully connected layers, where the first two layers have "Elu" activation and the last one has "softmax" activation function. The proposed novel architecture is shown in detail in Fig. 7.1.

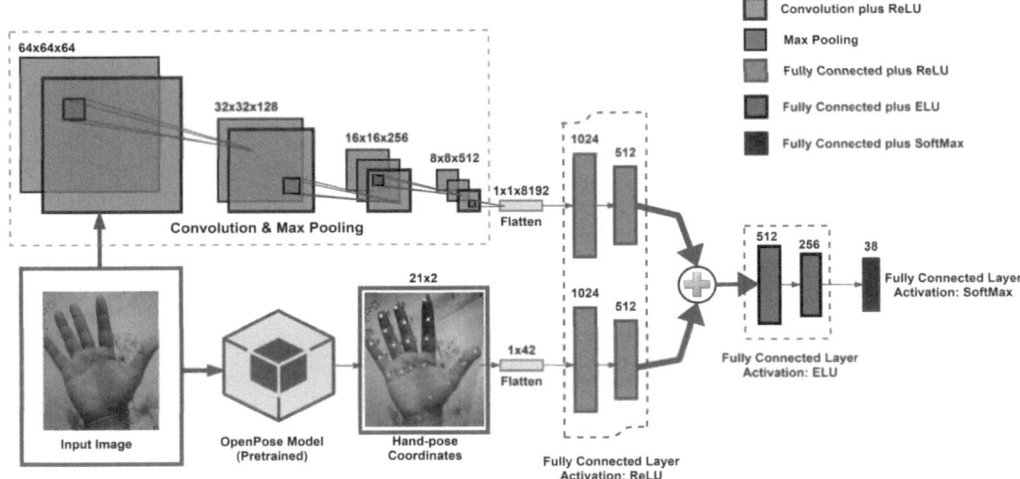

Figure 7.1 Concatenated BdSL Network

7.3.3 Training Method

Before the start of training, the two inputs are preprocessed as mentioned earlier. The preprocessed input of the image network is converted into a numpy file and fed directly into our proposed model. However, the preprocessed input for the pose estimation network has to be passed through the pretrained Openpose model first. The obtained outputs are hand pose co-ordinate values that are converted into another numpy file and fed as the second input of our proposed model. The labels associated with the images are also converted into another numpy file. The same procedure is followed at the beginning of the evaluation, except for the labels.

It is also mentioned that during training time, as a cost function, the cross entropy function is utilized. In order to reduce the cost function to a minimal value, a gradient descent based "Adam" optimization is used that has a learning rate of 0.001. The learning rate value is updated whenever the validation results of the model do not improve for 4 consecutive epochs using the reduced learning rate on plateau mechanism. For training, the weights are also initialized with small numbers. In order to avoid overfitting during training, an early stopping mechanism is used to interrupt training after a certain time if we see that the validation loss has not improved up to a certain number of epochs.

Table 7.1 Comparison of classification accuracy and used resources between the novel Concatenated BdSL Network and other existing work with the same dataset

	Models	
Criteria	Concatenated BdSL Network	Modified VGG-19 Image Network
Training	98.67%	97.68%
Validation	95.28%	91.52%
Testing	91.51%	89.6%
Image Size	64*64 pixels	224*224 pixels
Epoch	30	100
GPU	No	Yes

7.4 RESULTS

A detailed analysis of the performance of the model, along with the dataset and experimental setup, are discussed in this section.

7.4.1 Dataset and Experimental Setup

For this task, the Bengali Sign Language Dataset is used, which was collected by Rafi et al. [1] from Bangladesh National Federation of the Deaf (BNFD). It contains 11061 still images for training and 1520 for testing. There are 38 labels and the dataset is split into training and validation sets with 9959 and 1102 images, respectively, following previous works [1]. The experiments have been conducted in identical computational environments: - Framework: *Tensorflow 2.2*, Platform: *Google Colaboratory having a 1-core allocated Intel Xeon processor with 2.2GHz and 12.72GB RAM*, GPU: *None*.

7.4.2 Performance of Concatenated BdSL Network

As mentioned earlier, most existing BdSL recognition models only extract features from CNNs, which is called an "Image Network" in this proposed architecture. In addition to this, the proposed "Concatenated BdSL Network" also extracts hand keypoint values from the images. As a result, to analyse the effectiveness of the additional features, a comparison is made between the proposed model with respect to existing CNN based models like the modified VGG-19 Image Network [1] using the same dataset. Only the "Image Network" part of this proposed architecture has also been trained separately and can be considered as another CNN based model. This model is referred to as the "Only Image Network Model" in this chapter and is used as the second model for comparison with the Concatenated BdSL Network. Both the "Only Image Network Model" and the Concatenated BdSL Network have been trained for 30 epochs due to a lack of computational resources like a GPU.

The proposed Concatenated BdSL Network architecture achieved a validation score of 98.28 and overall test score of 91.51, whereas our "Only Image Network Model" got only 90 as a test score. Another comparison with the modified VGG-19

Image Network is given in Table. 10.1. The confusion matrix of the Concatenated BdSL Network is also included in Fig. 7.2.

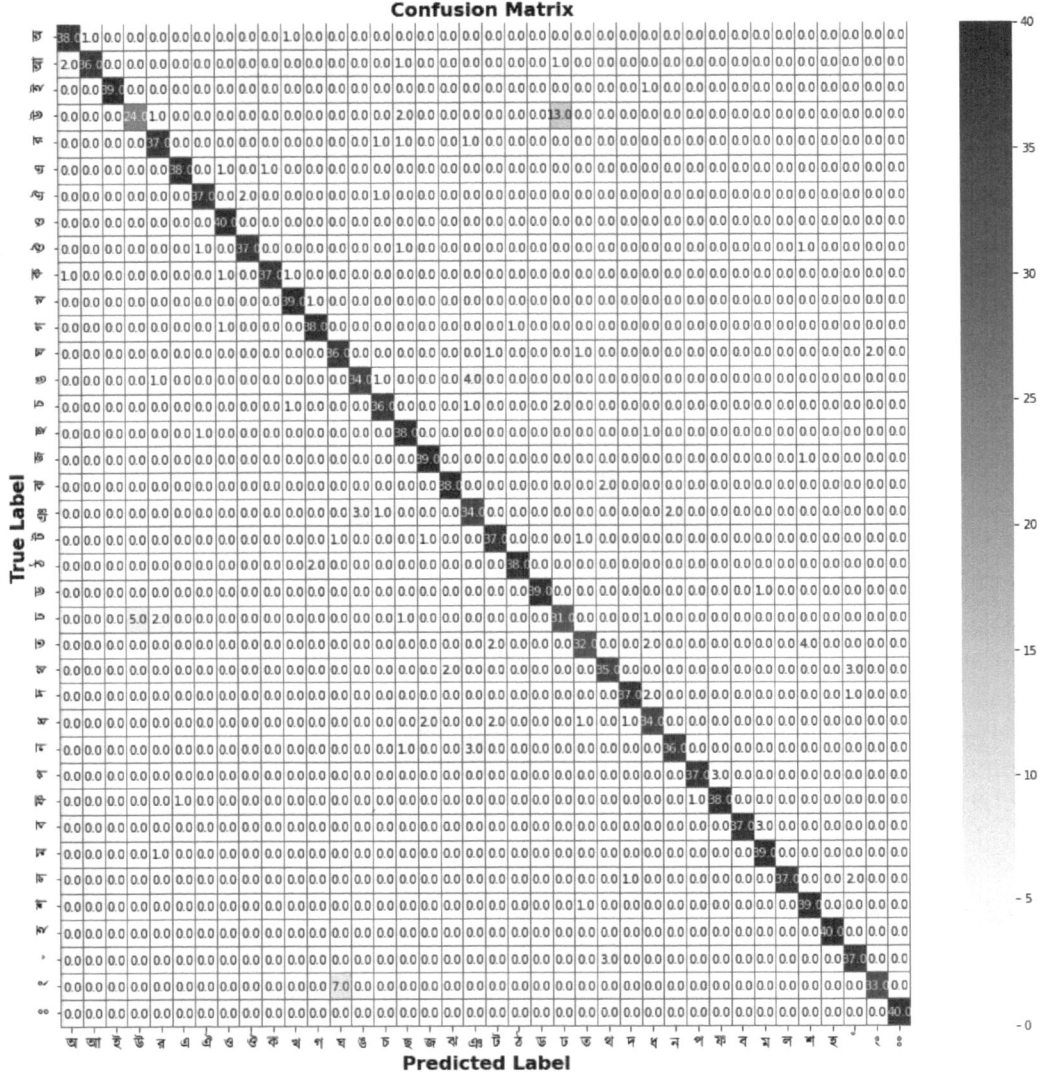

Figure 7.2 Confusion Matrix of the BdSL alphabet

7.5 DISCUSSION AND FUTURE SCOPE

There are 38 symbols in single-hand Bangla sign language and some of the alphabet symbols have very subtle differences with respect to others, which makes the posture of hands very important for recognising BdSL alphabets. With this view in mind, one of the highlights of the proposed method is to extract additional hand keypoint features from the pose estimation network to deal with the difficulty of recognising Bangla sign language. By comparing with deep learning based models that only extract visual features from CNNs for Bangla sign language, the significance of hand pose estimation features is analysed. Even though the modified VGG-19 Image Network was trained on higher-resolution images and for higher numbers of epochs, the Concatenated BdSL Network shows more promising results. It is also compared with the "Only Image Network Model," which is trained on the Image Network part of the proposed architecture only, excluding the Pose Estimation Network, and the comparison graph can be seen from Fig. 7.3. This is done to get a better understanding of the effectiveness of extracting hand keypoints.

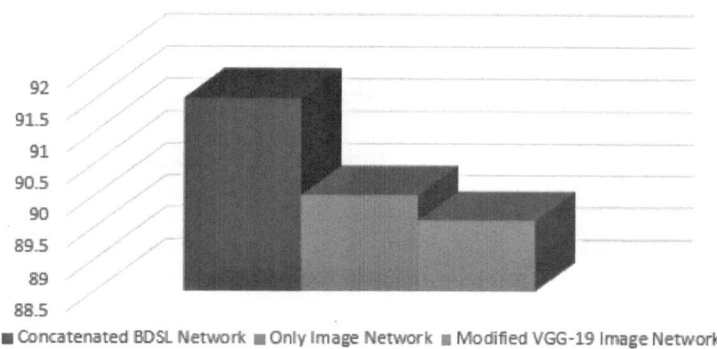

■ Concatenated BDSL Network ■ Only Image Network ■ Modified VGG-19 Image Network

Figure 7.3 Comparison of accuracy scores in test set among the different experimental models along with the existing work

The intuition is that the pose estimation network can estimate the relative position of a certain finger and its pose from the hand keypoint values because it assigns each finger joint of the hand a unique ID. This is an important factor in BdSL recognition to differentiate between two symbols that have very little difference. As a result, the Concatenated BdSL Network in addition to visual feature extraction from the image network, can also extract features to estimate finger positions at the same time. So, combining both visual and positional information helps the novel Concatenated BdSL Network achieve higher precision. The proposed novel model achieved state-of-the-art results on this Bengali Sign Language Dataset. However, further experiments could not be conducted with other similar works due to use of different datasets, like two-handed Bangla sign language [3, 8], very low training images [9], etc. Despite being a very good classification model, the confusion matrix is not 100 percent diagonal as some misclassifications are still prevalent.

For general people, even with human eyes, some signs are indistinguishable. They appear to be the same in terms of looks. The signs suffered mostly from recognizing these two symbols – 'Rosshu' and 'Dho'. From the confusion matrix, it is also seen that the Concatenated BdSL predicted 'Dho' instead of 'Rosshu' 13 times out of 40 cases. Other signs that suffered bad predictions are 'Umo'-'Eyo', 'Ta'-'Talebosh Sho'. From Fig. 7.4 it is seen that these six hand gestures are nearly identical to each other. So, there is still plenty of scope for improvement. If a pose estimation network could have been trained from scratch specifically for Bangla sign language hand pose

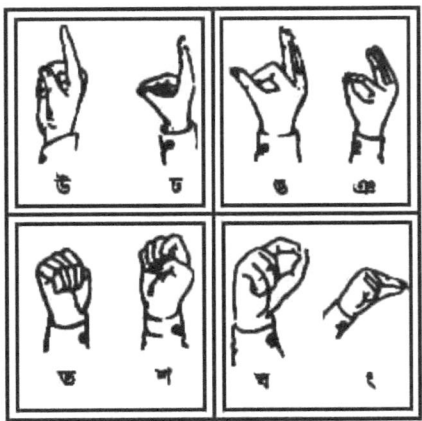

Figure 7.4 Nearly identical signs that are difficult to distinguish in BdSL

estimation, these issues could have been further addressed. However, this requires computational resources and a huge dataset that is beyond the scope of this work.

Collecting a Bangla sign language dataset, a pose estimation network can be trained from scratch in future work. With further work on the subject, the model can be developed to recognize real-time continuous sign language. It is because only by real-time implementation on continuous sign language, the tool can be made practically realisable. Finally, this will also help the model be suitable for commercial use.

Bibliography

[1] Rafi, A., Nawal, N., Bayev, N., Nima, L., Shahnaz, C. & Fattah, S. Image-based Bengali Sign Language Alphabet Recognition for Deaf and Dumb Community. *2019 IEEE Global Humanitarian Technology Conference (GHTC)*. pp. 1-7 (2019)

[2] Islam, S., Mousumi, S., Rabby, A., Hossain, S. & Abujar, S. A potent model to recognize Bangla sign language digits using convolutional neural network. *Procedia Computer Science*. **143** pp. 611-618 (2018)

[3] Hoque, O., Jubair, M., Islam, M., Akash, A. & Paulson, A. Real-time Bangladeshi sign language detection using faster R-CNN. *2018 International Conference on Innovation in Engineering and Technology (ICIET)*. pp. 1-6 (2018)

[4] Shanta, S., Anwar, S. & Kabir, M. Bangla sign language detection using sift and CNN. *2018 9th International Conference on Computing, Communication and Networking Technologies (ICCCNT)*. pp. 1-6 (2018)

[5] Oliveira, V. & Conci, A. Skin Detection using HSV color space. *H. Pedrini, & J. Marques De Carvalho, Workshops of Sibgrapi*. pp. 1-2 (2009)

[6] Zarit, B. & Boaz, J. Super, and Francis KH Quek. Comparison of Five Color Models in Skin Pixel Classification. Recognition, Analysis, and Tracking of Faces and Gestures in Real-Time Systems, 1999. *Proceedings. International Workshop On. IEEE.* (1999)

[7] Kuznetsova, A., Leal-Taixe, L. & Rosenhahn, B. Real-time sign language recognition using a consumer depth camera. *Proceedings of the IEEE International Conference on Computer Vision Workshops.* pp. 83-90 (2013)

[8] Islam, M., Rahman, M., Rahman, M., Arifuzzaman, M., Sassi, R. & Aktaruzzaman, M. Recognition Bangla Sign Language Using Convolutional Neural Network.

[9] Hossen, M., Govindaiah, A., Sultana, S. & Bhuiyan, A. Bengali sign language recognition using deep convolutional neural network. *2018 Joint 7th International Conference on Informatics, Electronics & Vision (ICIEV) and 2018 2nd International Conference on Imaging, Vision & Pattern Recognition (icIVPR).* pp. 369-373 (2018)

[10] Starner, T., Weaver, J. & Pentland, A. Real-time American sign language recognition using desk and wearable computer based video. *IEEE Transactions on Pattern Analysis and Machine Intelligence.* **20**, 1371-1375 (1998)

[11] Zafrulla, Z., Brashear, H., Starner, T., Hamilton, H. & Presti, P. American sign language recognition with the Kinect. *Proceedings of the 13th International Conference on Multimodal Interfaces.* pp. 279-286 (2011)

[12] Starner, T. & Pentland, A. Real-time American sign language recognition from video using hidden Markov models. *Motion-based Recognition.* pp. 227-243 (1997)

[13] Vogler, C. & Metaxas, D. Parallel hidden Markov models for American sign language recognition. *Proceedings of the Seventh IEEE International Conference on Computer Vision.* **1** pp. 116-122 (1999)

[14] Rajam, P. & Balakrishnan, G. Real-time Indian sign language recognition system to aid deaf-dumb people. *2011 IEEE 13th International Conference on Communication Technology.* pp. 737-742 (2011)

[15] Singha, J. & Das, K. Indian sign language recognition using eigen value weighted Euclidean distance based classification technique. *ArXiv Preprint ArXiv:1303.0634.* (2013)

[16] Imagawa, K., Lu, S. & Igi, S. Color-based hands tracking system for sign language recognition. *Proceedings Third IEEE International Conference on Automatic Face and Gesture Recognition.* pp. 462-467 (1998)

[17] Tanibata, N., Shimada, N. & Shirai, Y. Extraction of hand features for recognition of sign language words. *International Conference on Vision Interface.* pp. 391-398 (2002)

[18] Wang, C., Gao, W. & Shan, S. An approach based on phonemes to large vocabulary Chinese sign language recognition. *Proceedings of Fifth IEEE International Conference on Automatic Face Gesture Recognition.* pp. 411-416 (2002)

[19] Li, Y., Chen, X., Zhang, X., Wang, K. & Wang, Z. A sign-component-based framework for Chinese sign language recognition using accelerometer and sEMG data. *IEEE Transactions on Biomedical Engineering.* **59**, 2695-2704 (2012)

[20] Hasan, M., Sajib, T. & Dey, M. A machine learning based approach for the detection and recognition of Bangla sign language, *2016 International Conference on Medical Engineering, Health Informatics And Technology (MediTec).* pp. 1-5 (2016)

[21] Rahaman, M., Jasim, M., Ali, M. & Hasanuzzaman, M. Real-time computer vision-based Bengali sign language recognition. *2014 17th International Conference on Computer and Information Technology (ICCIT).* pp. 192-197 (2014)

CHAPTER 8

ChestXRNet: A Multi-class Deep Convolutional Neural Network for Detecting Abnormalities in Chest X-Ray Images

Ahmad Sabbir Chowdhury

Computer Science and Engineering, Chittagong Independent University, Chittagong, Bangladesh

Aseef Iqbal

Computer Science and Engineering, Chittagong Independent University, Chittagong, Bangladesh

CONTENTS

C Hest x-ray images are used as a primary diagnostic tool for different thoracic diseases like Pneumonia, Covid-19, SARS, Pneumothorax, etc. The abnormalities in these chest x-ray images are sometimes subtle and require expert eyes to identify. In this chapter, we propose a deep learning-based tool that automatically detects abnormalities in different chest radiography images and is able to identify with the highest probability if the patient is suffering from Covid-19, Effusions, Infiltration, Pneumonia, or Pneumothorax. A Convolutional Neural Network (CNN) based architecture named ChestXRNet, a supervised multi-class classification technique, is proposed to classify different thoracic chest x-ray images. The training dataset was compiled by collecting samples from three different open-source databases. We only scraped the necessary data for this project and built a customized dataset consists of 16,200 images. The performance of the proposed ChestXRNet model is compared with some pre-trained CNN models like DenseNet201, EfficientNetB7, and VGG16 for benchmarking using the same dataset. Our model acquired a training evaluation accuracy of 92.9% and a testing evaluation (prediction) accuracy of 82.8%. Model weights were saved and then integrated into a Flask based web server for a proper web application experience.

8.1 INTRODUCTION

In this day and age, to diagnose any kind of illness or abnormalities in our human body, diagnostic scans such as a CT scan or an X-ray is a must. Especially with the help of x-ray images, medical experts can examine and identify a wide variety of diagnoses, such as bone fracture, cancer, lung infections, etc. X-ray imaging can primarily be classified into two categories: hard X-rays and soft X-rays. Among these, hard x-rays have high photon energies and are widely used for medical purposes.

The Convolutional Neural Network (CNN) is a popular method for automated processing of medical images (x-rays, CT scans, MRI, etc.) for classification and localization of common thoracic diseases [1]. A well-trained radiologist may sometimes miss significant findings in x-rays as it is very challenging to distinguish between subtle but significant differences in visual features presented by different diseases. In this chapter, we proposed an automated chest x-ray abnormalities detection tool that can be used to ease and support radiologists by minimizing the detection and analyzing time. This tool is capable of classifying different thoracic diseases with the highest probability. This tool can be used in hospitals and diagnostic centers to reduce the patient waiting time, where a huge number of x-rays are analyzed every day. Additionally, in emergency situations, it can help speed up a diagnostic workflow.

This chapter used a Convolutional Neural Networks (CNNs) based architecture, named ChestXRNet, to classify different chest x-ray images. For building the deep learning architecture, we used Keras, the deep learning framework and the Tensor-Flow backend. We trained our model on more than 16,000 chest x-ray images and only used the posteroanterior (PA) view of the chest x-ray images. We created a customized dataset where 6 different classes of chest x-ray images were labelled and stored. The six different classes are Covid-19, Effusions, Infiltration, Pneumonia, Pneumothorax, and No-Findings.

8.2 RELATED WORK

Recent technological advances showed that using different Deep Learning based approaches can outperform experts when it comes to diagnosing medical images [2, 3]. Among the existing deep neural network architectures, the CNN performs best for detecting any kind of abnormalities or fault in medical scan images. CNNs were developed taking images into consideration, so they have a high level of accuracy when it comes to feature extraction of medical images [4].

For chest x-ray interpretation, we can use different CNN methods, such as binary classification, multi-label classification, multi-class classification [5, 6, 7, 8]. Among all, binary and multi-label classification are the two most common methods and, in this literature, all the proposed approaches have used these two methods. For instance, in [9] a model called ChestNet was proposed, which consists of a classification branch and an attention branch for the diagnosis of different thorax diseases on the chest radiograph. This model was also evaluated against three state-of-the-art deep learning models using 14 different chest x-ray datasets [10].

Chest x-ray images also played a vital role in diagnosing Covid-19 patients at the beginning of the pandemic. In [7], a model was proposed to classify Covid-19 from chest x-ray images using transfer learning [11] and binary classification method. This model [7] can deal with irregularities in the dataset by investigating different boundaries using a class decomposition method. Currently, we know the best screening method used to detect Covid-19 is Polymerase Chain Reaction (PCR) [8] testing, which can detect SARS- CoV-2 RNA from respiratory specimens. Although PCR testing is very standard and highly sensitive, it is also very complicated and time-consuming. Thus, the best alternative method to detect Covid-19 is chest radiography. In [12] the proposed model makes predictions using an explainability method to gain deeper insights into critical factors associated with Covid cases. This network architecture makes heavy use of a lightweight residual projection-expansion-projection-extension (PEPX) design pattern.

Chest radiography images have also played a huge role in detecting pneumonia using different neural network techniques. In [2], a 121-layer convolutional neural network was trained on over 100,00 frontal view x-ray images of 14 diseases. This model performance was compared with four practicing radiologists, and the model outperformed all the radiologists when it came to the F1 score. The proposed model was a pre-trained DenseNet model, and it used a binary classification method to successfully detect different pneumonia-positive patients.

In our study, we took a different approach than the ones discussed earlier. We used a multi-class classification technique to identify each disease class separately. The problem with multi-label classification is that it overlaps classes as many chest x-ray images have similar features. Our approach is different, as all our data were predefined, pre-labelled and stored in different datasets, which helped our model learn all the different features smoothly. Unlike other literature, we used Label Smoothing

Regularization (LSR) [13] and different call-back functions to improve our model training process.

8.3 METHODOLOGY

The overall system architecture of the proposed method is shown in Figure 8.1. To build our proposed model, the chest x-ray images were collected from three different databases. From the NIH Chest X-Ray database [14, 24], we collected "Effusions", "Infiltration", "Pneumonia" and "Pneumothorax" affected chest x-ray data. From the Covid-19 Chest X-Ray database [15], we only collected the Covid-positive radiography images, and finally, we collected some normal chest x-ray images for comparison purposes.

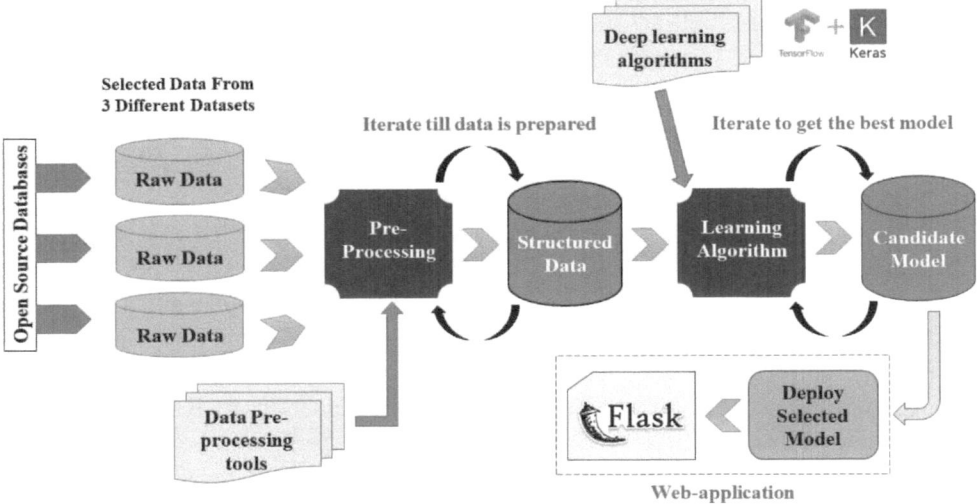

Figure 8.1   Overall system architecture of the ChestXRNet classifier model.

Among all the different chest x-ray images, there were a lot of similarities. That is why it was important to extract all the different features of all those different classes. The next step in our proposed system architecture was to pre-process all the collected data using different pre-processing tools. After the pre-processing, we fed all the data into our proposed ChestXRNet model, which was built using Keras and TensorFlow backend. We fine-tuned different hyperparameters several times so that our model achieved the best results. After the model training and evaluation part, the model weights were saved so that they can be integrated into the Flask server for deployment purposes.

8.3.1   Data Preprocessing

First, we have to check whether all our input data is of the same size or not. If the input variables are not scaled properly, they may result in an unstable learning process. Data preparation involves using techniques such as normalization and standardization, to rescale input and output variables prior to training a CNN model. Based on data inspection, images are scaled to a size of 64 by 64 and normalized to values (0,1). This process will help our model learn more steadily, as now all the image data are the same size (64,64). As we are doing multi-class classification, our chest x-ray image dataset contains categorical values, but the Machine Learning/Deep Learning algorithms work better with numerical inputs. That's why all the categorical values need to be converted to numerical data.

8.3.2 Data Augmentation

The x-ray image data which we are looking for has to have good diversity in terms of varying sizes, lighting conditions and correct poses. The dataset prepared for this project contains only 16,200 images, which is not very ideal if we want to make our deep learning model work perfectly. This will cause our model to overfit, but fortunately, we can easily overcome this problem by using data augmentation techniques [16]. In data augmentation, we can generate a lot of new data with the help of our existing data, which will eliminate the problem of having fewer data. Different data augmentation techniques which we used for our proposed model are: rescale, zoom-range, shear-range, rotation-range, width-shift-range, horizontal-flip, height-shift-range.

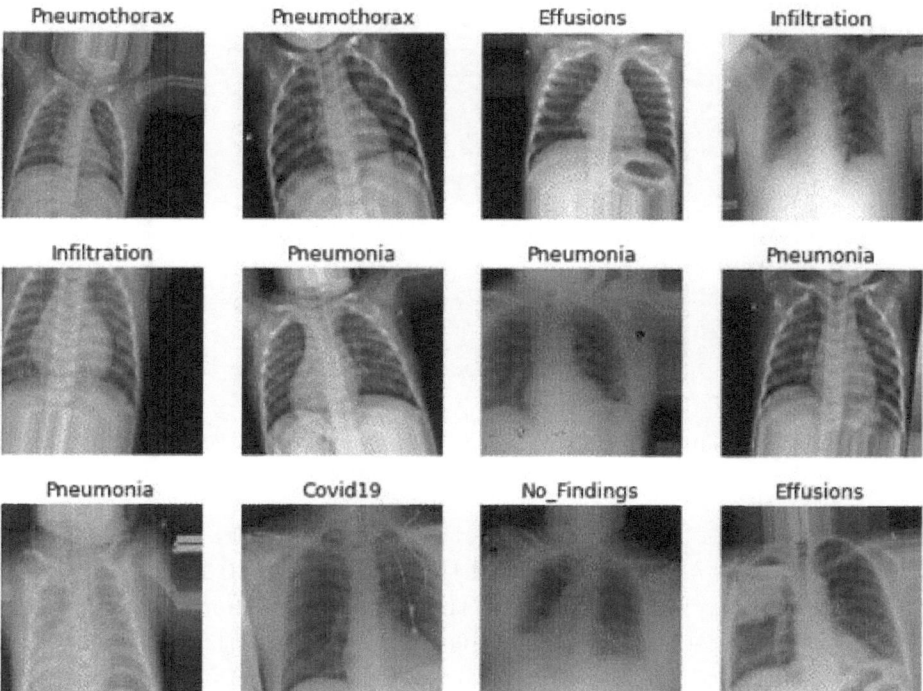

Figure 8.2 Training x-ray images after applying augmentation.

8.3.3 Proposed ChestXRNet Model

In the proposed architecture, we first instantiated our sequential model object, and then we added layers to it one by one. We added a 2D convolutional layer to process the 2D chest x-ray input images. From Figure 8.3 we can see the first argument passed to the Conv2D layer function is the number of output channels. We have 32 output channels in the first convolutional layer. The next input is the kernel size. We have chosen a 3x3 moving window, followed by the strides (1, 1). Next, the activation function is a rectified linear unit (ReLU) and finally, the input size was fixed to 64x64. Next, we added a 2D max-pooling layer to shrink the input size. After that, we used a second Conv2D layer, used 32 output channels, and again added a 2D max-pooling layer. Finally, we added the last and final Conv2D layer plus a max-pooling layer, with again 32 output channels. After building all four convolutional hidden layers, we flattened the output and entered it into the fully connected layers. In the first Dense layer, there are 128 output channels, but the last Dense layer has only 6 output channels, as we are classifying between 6 different classes. Also, in the last Dense layer, we used the SoftMax activation function because of the categorical class mode.

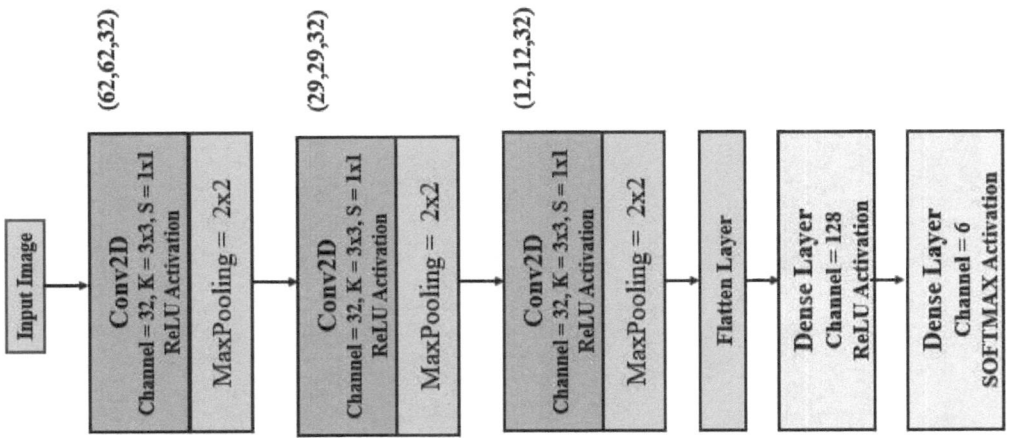

Figure 8.3 Schematic of the proposed ChestXRNet model architecture for input shape (64,64).

In Figure 8.3 we can see that we used three convolutional layers, three pooling layers and two dense layers. After building the ChestXRNet model, we got a total of 167,750 trainable parameters. These trainable parameters are basically the total number of trainable elements that are affected by backpropagation in our model. Among 167,750 trainable parameters, 148358 parameters come from the output layers.

Different techniques and fine-tuning methods are being used to improve the training result of the proposed model. We downscaled the images to 64 x 64 and normalized them based on the mean and standard deviation before putting them in the model for better accuracy. While training, we used a framework called 'wandb', meaning

Weights and Biases. It helped us save our hyperparameters and output metrics. For optimizing our model, we used the 'Adam' [17] optimizer with a learning rate of 0.0001.

8.3.4 Proposed Transfer Learning Methods for Benchmarking

To observe how our ChestXRNet model is performing, we compared the results with three pre-trained models. They are Densenet201 [18], EfficientNetB7 [19] and VGG 16 [20]. For image classification, we froze the early convolutional layers of the pre-trained networks and only trained the last few layers that make a prediction. The availability of annotated medical images is limited and because of that the classifications of medical images especially radiography images remain one of the biggest challenges in medical diagnosis. One way to overcome the challenge is using Transfer Learning techniques [21]. Large-scale image datasets (such as ImageNet) provide effective solutions for models that are using a small-scale dataset.

DenseNet-201 is a 201-layers deep convolutional neural network. Total parameters of DenseNet201 are 19,428,358 but only 1,106,374 were trainable parameters. EfficientNet is built for ImageNet classification and contains 1000 classes of labels. As we do not need most of the top layers of this model, they can be excluded while loading the model [19]. Total parameters of EfficientNetB7 are 65,572,694 but only 1,475,014 were trainable parameters. VGG16 is a convolutional neural network that is 16 layers deep. We can load a pre-trained version of the network trained on more than a million images from the ImageNet database [20]. Total parameters of VGG16 are 15,313,286 but only 598,598 were trainable parameters. As mentioned earlier, trainable parameters are basically the total number of trainable elements that are affected by backpropagation in our model. Certain parameters are non-trainable because we cannot optimize the values with our training data. We usually come across non-trainable parameters during transfer learning. For all the three pre-trained models, the input size was fixed to 224x224, batch size was fixed to 64, and the learning rate was 0.0001 for all models. Also, a minimum of 40 epochs and an Adam optimizer were used during all model training.

8.3.5 Callbacks in Keras

While training a CNN model, too many epochs may sometimes lead to overfitting the whole model. On the other hand, very few epochs may result in an underfit model. We can overcome this problem by using the 'EarlyStopping' function [22]. This method helps us to stop training when it sees the model performance is not improving on a holdout validation dataset. The Keras API called "callbacks" is used for adding the EarlyStopping function in our model while training.

When the model is in training mode, sometimes we noticed that the performance was not improving and the validation accuracy level was not changing. To counter that problem, we used the 'ReduceLROnPlateau' function. It automatically reduces the learning rate and starts training our model again. As a result, we see a boost in our model performance level and at the same time, we see that training loss starts decreasing.

8.4   RESULT ANALYSIS AND DISCUSSION

In this section, the performance of the proposed schemes are presented with a visual interpretation. The dataset was divided into an 80%-20% (training and validation) ratio. In the training folder there were 13,500 images, in the validation folder there were 2700 images, and in the test folder there were 600 images.

8.4.1   Data Description and Datasets

For building the model, we used in total 16,200 images and a separate 600 images just for testing the model. A total of three datasets were used in this project: the NIH Chest X-Ray Dataset [14, 24], the Covid-19 Chest X-Ray Dataset [15], and the Normal Chest X-Ray Image Dataset [23]. From the 'NIH Chest X-Ray Dataset', 4 different classes were extracted. They are Effusions (3000 Images), Infiltration (3000 Images), Pneumonia (3000 Images) and Pneumothorax (3000 Images). This dataset was mined from the associated radiological reports using NLP. The dimensions of the image samples are 224x224. From the 'Covid-19 Chest X-Ray Dataset', only Covid Positive (1200 Images) chest x-rays were taken, and from the 'Normal Chest X-Ray Dataset' we extracted the No-Findings (3000 Images) class data.

8.4.2   Experimental Setup

The entire research work was performed in two separate setups. To build and train some of our models, we used the Google Colab environment. To integrate the model in our Flask based web application and further advance model training, we used a CUDA core-enabled computer. Numerous classification tasks are used for evaluating the performance of the proposed CNN model, such as accuracy, precision, recall and F1 score. All these evaluations were also done in two different system setups. Other than that, the tools/frameworks that are used to build the proposed model are the OS platform Windows 10, Python 3.7, TensorFlow 2.3.1, TensorFlow-GPU 2.3.1, Keras 2.4.3, OpenCV 4.5.0, Flask 1.1.2, Scikit-Learn, HTML, CSS, and Bootstrap. The specifications of the CUDA core-enabled test computer are MSI GTX 1650 Max-Q, 1024 Cores, CUDA 7.5, 4GB GDDR6 VRAM, Intel Core i5 9300H @2.40GHz (4 Core, 8 Threads), 8 GB DDR4 RAM, and 512 SSD.

8.4.3   ChestXRNet Model's Training, Validation Accuracy and Loss

We used 70 epochs while training the proposed model. After the training was done, we saw a 0.93 training accuracy score and a 0.86 validation accuracy score. Also, from Figure 8.4 we can see the training loss (0.183) is less than the validation loss (0.445).

In plotting Figure 8.5(a), we can see that in some moments our training accuracy was not improving, but after some epochs, we saw a boost in our accuracy graph. This happened because of using the 'ReduceLROnPlateau' function. This function helps the learning rate change automatically when it sees the model performance is not improving and the training-validation loss is not decreasing enough. This results in some saturation around the training and validation accuracy learning curve.

Training Evaluation

```
print("Model evalution result --> Training Data")
model.evaluate(train_generator)

Model evalution result --> Training Data
422/422 [==============================] - 161s 383ms/step - loss: 0.1838 - accuracy: 0.9291

[0.18382394313812256, 0.929111123085022]
```

Validation Evaluation

```
print("Model evalution result --> Validation Data")
model.evaluate(validation_generator)

Model evalution result --> Validation Data
85/85 [==============================] - 30s 354ms/step - loss: 0.4454 - accuracy: 0.8610

[0.4454165995121002, 0.8609981536865234]
```

Figure 8.4 Training and Validation evaluation scores.

Also, one of the main reasons why the learning curve of validation accuracy is a little noisy is because the validation dataset has fewer examples as compared to the training dataset. In plotting Figure 8.5(b), we can see the training and validation loss continues to decrease throughout all 70 epochs. On observation, we found the training loss is never excessive, indicating that our model didn't face any overfitting or underfitting issues.

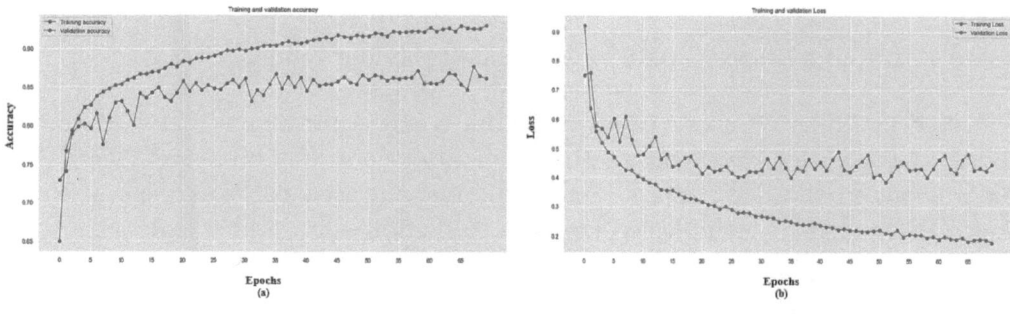

Figure 8.5 Plotting of training, validation accuracy and loss (ChestXRNet Model).

8.4.4 Result Comparison Between ChestXRNet and Other Pre-Trained Models

While training DenseNet201, EfficientNetB7 and VGG16 models, we used Label Smoothing Regularization (LSR) and also the callback functions ReduceLROn-Plateau and EarlyStopping. These functions helped us get better results and stopped the training process earlier than our initial epochs when there is no further improvement in our model training. Similar phenomena occurred while training the DenseNet201 and EfficientNetB7 models. Both the models stopped training midway because no further improvements were occurring. So, after 55 epochs, DenseNet201 achieved a 0.90 training accuracy score and a 0.85 validation accuracy score. From all

4 models, EfficientNetB7 achieved the lowest training accuracy (0.75) and validation score (0.69). Lastly, the VGG16 achieved a 0.86 training score and a 0.82 validation score after 50 epochs. It was a perfectly fitted model where we didn't see any major saturation between the training and validation learning curves. In Figure 8.6, we can see the training accuracy plot comparison of all 4 models. Among all, our proposed ChestXRNet model performed the best. It completed all 70 epochs and it achieved a training accuracy score of 0.93 and validation accuracy score of 0.86, which is the highest among all four models.

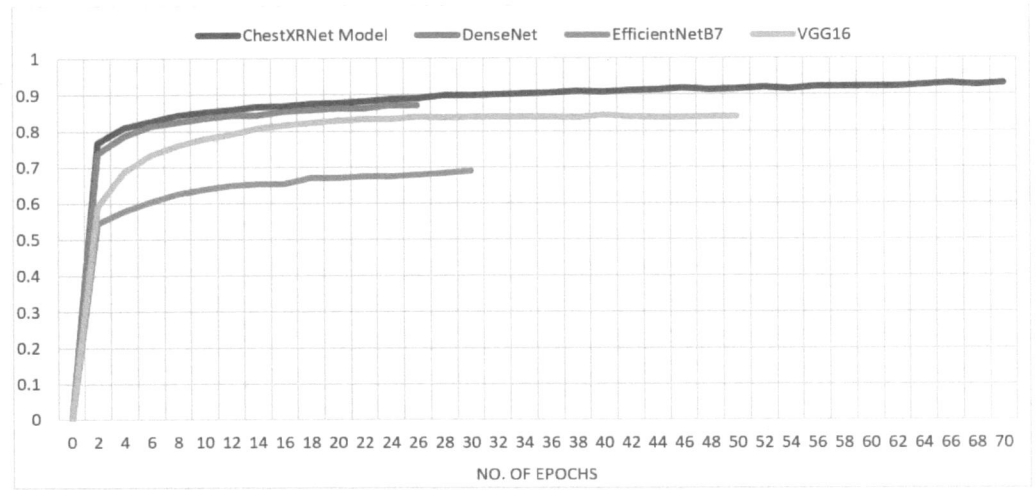

Figure 8.6 Training accuracy plot comparison of the four models.

Below in Table 8.1, we can see all the result comparisons between the four models. Again, when it comes to the prediction score, our proposed ChestXRNet model achieved the best results. For prediction, we used 623 test images and our ChestXRNet model accomplished the best prediction score, which is 0.8282.

Table 8.1 Four CNN models result comparison.

	Accuracy	Loss	Val_acc	Val_loss	Prediction
ChestXRNet	0.929	0.1838	0.8609	0.4454	0.8282
DenseNet201	0.905	0.6447	0.8547	0.7321	0.8218
EfficientNetB7	0.755	0.9236	0.6994	1.0189	0.6773
VGG16	0.863	0.7188	0.8255	0.7923	0.7833

8.4.5 Model Evaluation and Prediction

In this section of the chapter, model evaluation and prediction of the proposed schemes are presented with visual interpretation. Our ChestXRNet model and the other three pre-trained models showed different prediction results. We tested all the models with unseen images, some predicted better and some didn't. We first scaled and normalized all the testing images for better results. To see how our model performed with unseen chest x-ray images, we used 'confusion matrix' and 'classification report' techniques. After the evaluation of the ChestXRNet model, it got a 0.83 prediction score and the misclass value was only 0.17. From Figure 8.7 we can see that the 'Pneumonia' prediction score is 0.91, which is the best result in our confusion matrix. If we compare the prediction scores of all six classes, we can see on average almost all the classes scored 0.80, which is excellent.

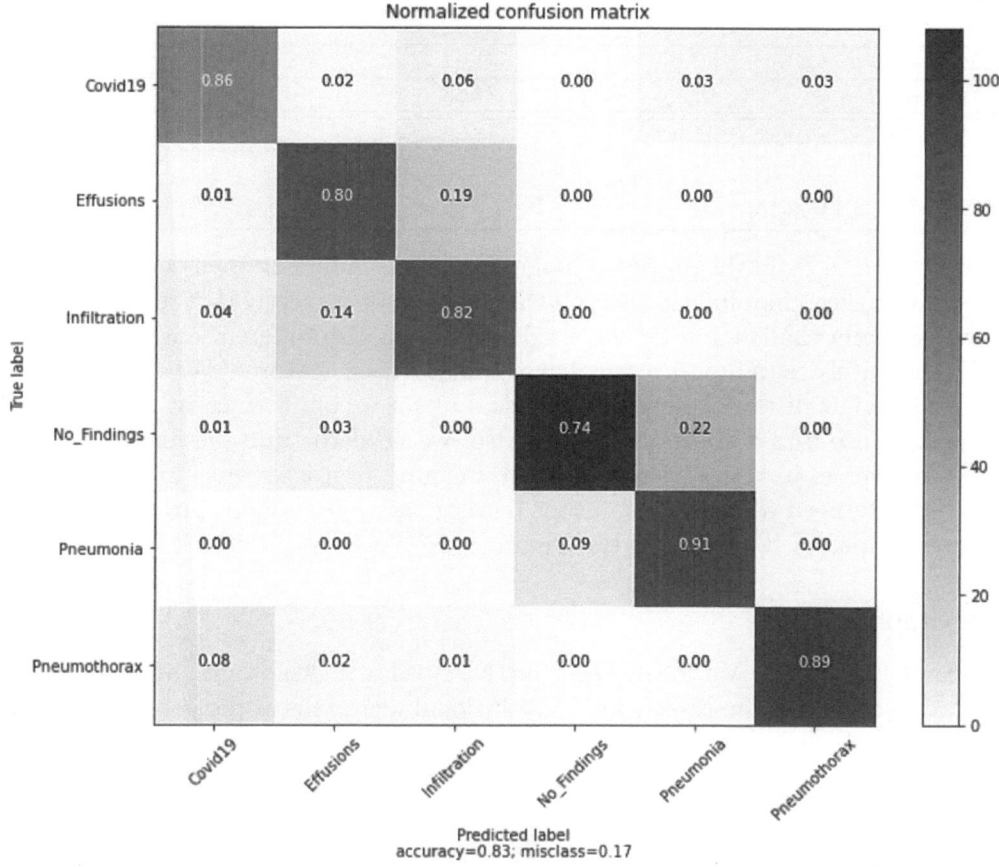

Figure 8.7 ChestXRNet Model's Confusion Matrix with prediction scores.

Of all three pre-trained models, the DenseNet201 model prediction score was the highest (0.82). The EfficientNetB7 model prediction score was the lowest among all the models (0.67). On the other hand, the VGG16 prediction score was 0.78. The misclass value of the DenseNet201 model was 0.17, the EfficientNetB7 model was 0.32 and the VGG16 model was 0.21. Other than calculating the average prediction score, we also calculated metrics such as 'Precision', 'Recall' and 'F1 Score' for all the models. In the classification report, Table 8.2, we can see all the classes' F1 scores. The rows define six different classes of our model, plus the accuracy, macro average and weighted average score. The columns show the model names and F1 scores.

Table 8.2 F1 score comparison of all four multi-class models.

	ChestXRNet	DenseNet201	EfficientNetB7	VGG16
Covid-19	0.81	0.78	0.49	0.72
Effusions	0.80	0.78	0.72	0.74
Infiltration	0.79	0.71	0.38	0.68
No-Findings	0.82	0.85	0.79	0.81
Pneumonia	0.81	0.85	0.69	0.77
Pneumothorax	0.93	0.91	0.74	0.91
Accuracy	0.83	0.82	0.68	0.78
Macro Avg	0.83	0.81	0.63	0.77
Weighted Avg	0.83	0.82	0.66	0.78

8.5 CONCLUSION

We took data to build the proposed model from three different open-source databases, so we have to keep in mind that not all the data will be correctly labelled and therefore it will be really challenging for our model to distinguish different chest x-ray images 100% accurately as many diseases have similar visual features. We only considered the frontal view of the chest x-ray images, but for a complete chest x-ray classifier, we need to use x-rays from all angles. Also, we need to train our model with data that are diverse, in terms of region, race, imaging protocols, etc. In the future, we will certainly need a larger and better dataset, as well as more powerful GPU and hardware capacity to store and train data.

Bibliography

[1] S. M. Anwar, M. Majid, A. Qayyum, M. Awais, M. Alnowami, and M. K. Khan, "Medical image analysis using convolutional neural networks: a review," *J. Med. Syst.*, vol. 42, no. 11, 2018, doi: 10.1007/s10916-018-1088-1.

[2] Rajpurkar, Pranav, Jeremy Irvin, Kaylie Zhu, Brandon Yang, Hershel Mehta, Tony Duan, Daisy Ding et al. "Chexnet: Radiologist-level pneumonia detection on chest x-rays with deep learning." *arXiv preprint arXiv:1711.05225 (2017)*.

[3] Rajpurkar, Pranav, Jeremy Irvin, Robyn L. Ball, Kaylie Zhu, Brandon Yang, Hershel Mehta, Tony Duan et al. "Deep learning for chest radiograph diagnosis: A retrospective comparison of the CheXNeXt algorithm to practicing radiologists." *PLoS medicine* 15, no. 11 (2018): e1002686.

[4] Khan, Hassan Ali, Wu Jue, Muhammad Mushtaq, and Muhammad Umer Mushtaq. "Brain tumor classification in MRI image using convolutional neural network." *Math. Biosci. Eng* 17, no.5 (2020): 6203-6216.

[5] Zhang, Min-Ling, and Zhi-Hua Zhou. "A review on multi-label learning algorithms." *IEEE Transactions on Knowledge and Data Engineering* 26, no. 8 (2013): 1819-1837.

[6] Mazurowski, Maciej A., Mateusz Buda, Ashirbani Saha, and Mustafa R. Bashir. "Deep learning in radiology: An overview of the concepts and a survey of the state of the art with focus on MRI." *Journal of Magnetic Resonance Imaging* 49, no. 4 (2019): 939-954.

[7] Asif, Sohaib, Yi Wenhui, Hou Jin, and Si Jinhai. "Classification of COVID-19 from Chest X-ray images using Deep Convolutional Neural Network." In *2020 IEEE 6th International Conference on Computer and Communications (ICCC)*, pp. 426-433. IEEE, 2020.

[8] Wang, Wenling, Yanli Xu, Ruqin Gao, Roujian Lu, Kai Han, Guizhen Wu, and Wenjie Tan. "Detection of SARS-CoV-2 in different types of clinical specimens." *JAMA* 323, no. 18 (2020): 1843-1844.

[9] Wang, Hongyu, and Yong Xia. "Chestnet: A deep neural network for classification of thoracic diseases on chest radiography." *arXiv preprint arXiv:1807.03058* (2018).

[10] Wang, Xiaosong, Yifan Peng, Le Lu, Zhiyong Lu, Mohammadhadi Bagheri, and Ronald M. Summers. "Chestx-ray8: Hospital-scale chest x-ray database and benchmarks on weakly-supervised classification and localization of common thorax diseases." In *Proceedings of the IEEE Conference on Computer Vision and Pattern Recognition*, pp. 2097-2106. 2017.

[11] Raghu, Maithra, Chiyuan Zhang, Jon Kleinberg, and Samy Bengio. "Transfusion: Understanding transfer learning for medical imaging." *Advances in Neural Information Processing Systems* 32 (2019).

[12] Wang, Linda, Zhong Qiu Lin, and Alexander Wong. "Covid-net: A tailored deep convolutional neural network design for detection of Covid-19 cases from chest x-ray images." *Scientific Reports* 10, no. 1 (2020): 1-12.

[13] Zhang, Chang-Bin, Peng-Tao Jiang, Qibin Hou, Yunchao Wei, Qi Han, Zhen Li, and Ming-Ming Cheng. "Delving deep into label smoothing." *IEEE Transactions on Image Processing* 30 (2021): 5984-5996.

[14] Wang, Xiaosong, Yifan Peng, Le Lu, Zhiyong Lu, M. Bagheri, and R. Summers. "Hospital-scale chest x-ray database and benchmarks on weakly-supervised classification and localization of common thorax diseases." In *IEEE CVPR*, vol. 7. 2017.

[15] Cohen, Joseph Paul, Paul Morrison, Lan Dao, Karsten Roth, Tim Q. Duong, and Marzyeh Ghassemi. "Covid-19 image data collection: Prospective predictions are the future." *arXiv preprint arXiv:2006.11988* (2020).

[16] Wong, Sebastien C., Adam Gatt, Victor Stamatescu, and Mark D. McDonnell. "Understanding data augmentation for classification: when to warp?." In *2016 International Conference on Digital Image Computing: Techniques and Applications (DICTA)*, pp. 1-6. IEEE, 2016.

[17] Kingma, D. P., and J. Lei Ba. "Adam: A method for stochastic optimization 3rd Int. Conf. Learn." *Representations (Preprint 1412.6980 v9)* (2015).

[18] Jaiswal, Aayush, Neha Gianchandani, Dilbag Singh, Vijay Kumar, and Manjit Kaur. "Classification of the COVID-19 infected patients using DenseNet201 based deep transfer learning." *Journal of Biomolecular Structure and Dynamics* (2020): 1-8.

[19] Tan, Mingxing, and Quoc Le. "EfficientNet: Rethinking model scaling for convolutional neural networks." In *International Conference on Machine Learning*, pp. 6105-6114. PMLR, 2019.

[20] Tammina, Srikanth. "Transfer learning using VGG-16 with deep convolutional neural network for classifying images." *International Journal of Scientific and Research Publications (IJSRP)* 9, no. 10 (2019): 143-150.

[21] Shie, Chuen-Kai, Chung-Hisang Chuang, Chun-Nan Chou, Meng-Hsi Wu, and Edward Y. Chang. "Transfer representation learning for medical image analysis." In *2015 37th Annual International Conference of the IEEE Engineering in Medicine and Biology Society (EMBC)*, pp. 711-714. IEEE, 2015.

[22] Wei, Yuting, Fanny Yang, and Martin J. Wainwright. "Early stopping for kernel boosting algorithms: A general analysis with localized complexities." *IEEE Transactions on Information Theory* 65, no. 10 (2019): 6685-6703.

[23] Kermany, Daniel, Kang Zhang, and Michael Goldbaum. "Labeled optical coherence tomography (oct) and chest X-ray images for classification." *Mendeley Data* 2, no. 2 (2018).

[24] R. (NIH/CC/DRD) [E] Summers, "NIH Chest X-Ray Dataset," National Institutes of Health - Clinical Center, 2017. https://nihcc.app.box.com/v/ChestXray-NIHCC.

CHAPTER 9

Achieving Human Level Performance on the Original Omniglot Challenge

Shamim Ibne Shahid

National University of Science and Technology MISiS, Moscow, Russia

CONTENTS

Eight years ago, since the Omniglot data was first released, very few papers have addressed the original Omniglot challenge, which is to carry out within-alphabet, one-shot classification tasks as opposed to selecting the test samples between the alphabets. Most researchers have made the task easier by introducing new splits in the dataset and have taken advantage of significant sample and class augmentation. Amongst the deep learning models that have adopted the Omniglot challenge as it is, the Recursive Cortical network has the highest performance of 92.75%. In this chapter, we introduce a new similarity function to aid in the training procedure of matching network, which helps achieve 95.75% classification accuracy on the Omniglot challenge without requiring any data augmentation.

9.1   INTRODUCTION

Unlike the conventional classification method where a model is explicitly trained with image categories that are also present in the test set, one-shot training process occurs with what is called in the literature a background set, and the image categories included here and the evaluation set are not the same. Rather than learning what features a certain category of image might consist of, one-shot training aims to force the model to understand what set of attributes could be more important in deciding whether an image is similar to another. In other words, the model is learning to learn the features in an image from an unknown distribution which are more responsible for distinguishing its category. This is why a model trained by following such a strategy can not only help in classifying unseen categories of samples, but can also be applied to perform one shot classification in a completely disjoint dataset, such as MNIST (Koch et al., 2015). However, a model trained this way with the Omniglot background set may not be suitable for doing such classification tasks on datasets containing object categories, because the attributes of an object are formed by not just different patterns of lines but also colors and textures. In this chapter, we will be limited to carrying out experiment on the Omniglot and MNIST datasets. In practice, the process of one-shot training has been extended to both image and language modeling. In the next section we discuss related works. In section 9.3, we briefly outline our approach. In section 9.4 and 9.5, the evaluation procedure on Omniglot and MNIST datasets are described, respectively. In the conclusion, we included the pros and cons of our method.

9.2 RELATED WORK

In terms of architecture, our model is similar to the Convolutional Siamese net (Koch et al., 2015), except with additional convolutional layer. We have followed the general training strategy outlined in Matching net (Vinyals et al., 2016), for one-shot image classification. Prototypical net (Snell et al., 2017) has generalized the one-shot setting introduced in matching net to few-shot learning by comparing the query embedding to the mean of support embeddings from each class and it also differs by the choice of distance metric. Our choice of the similarity function is influenced by the attribute label noise identification technique introduced in (J. Speth and E. M. Hand, 2019). Here, an attribute vector indicating the level of similarity of various attributes between two images passed through a Siamese net was obtained by the weighted combination of the distance between the corresponding feature maps. In our work, we assumed that the negative summation of the elements of such a vector is the measure of overall similarity between the two images. Recursive Cortical network (George et al., 2017) has introduced a generative vision model that can combine recognition, segmentation and reasoning for learning in a data efficient way. In the few-shot learning domain, the Prototype relational net (X. Liu et al., 2019) has exploited the use of a new loss function which takes into account both inter-class and intra-class distance. The learning process in Hierarchal Bayesian Program Learning (Lake, et al., 2013) is based on compositionality and causality. In Bayesian Program Learning (Lake et al., 2015), a concept is represented as simple programs explaining the observed data. Although the learning procedure of this algorithm is more similar to adults, the inductive bias acquired through the learning may help explain why some children struggle to learn some characters and also determine the best method to teach them.

9.3 METHODOLOGY

The query set for a one-shot learning scheme can be defined as $Q=\{(\{\mathbf{x_{ij}}\}_{j=1}^{q},\mathbf{y_i})\}_{i=1}^{k}$, where $\mathbf{y_i}$ denotes the one-hot representation of the category i and each $x \in \{\mathbf{x_{ij}}\}_{j=1}^{q}$ indicates a sample belonging to the class i. For each query$(\hat{\mathbf{x}}, \mathbf{t}) \in Q$ and the support set $S = (\mathbf{x_i}, \mathbf{y_i})_{i=1}^{k}$, we define the predicted label for the query $\hat{\mathbf{x}}$ as,

$$\hat{\mathbf{y}} = \sum_{i=1}^{k} \frac{e^{C(f(\mathbf{x_i}), f(\hat{\mathbf{x}}))}}{\sum_{j=1}^{k} e^{C(f(\mathbf{x_j}), f(\hat{\mathbf{x}}))}} \mathbf{y_i}$$

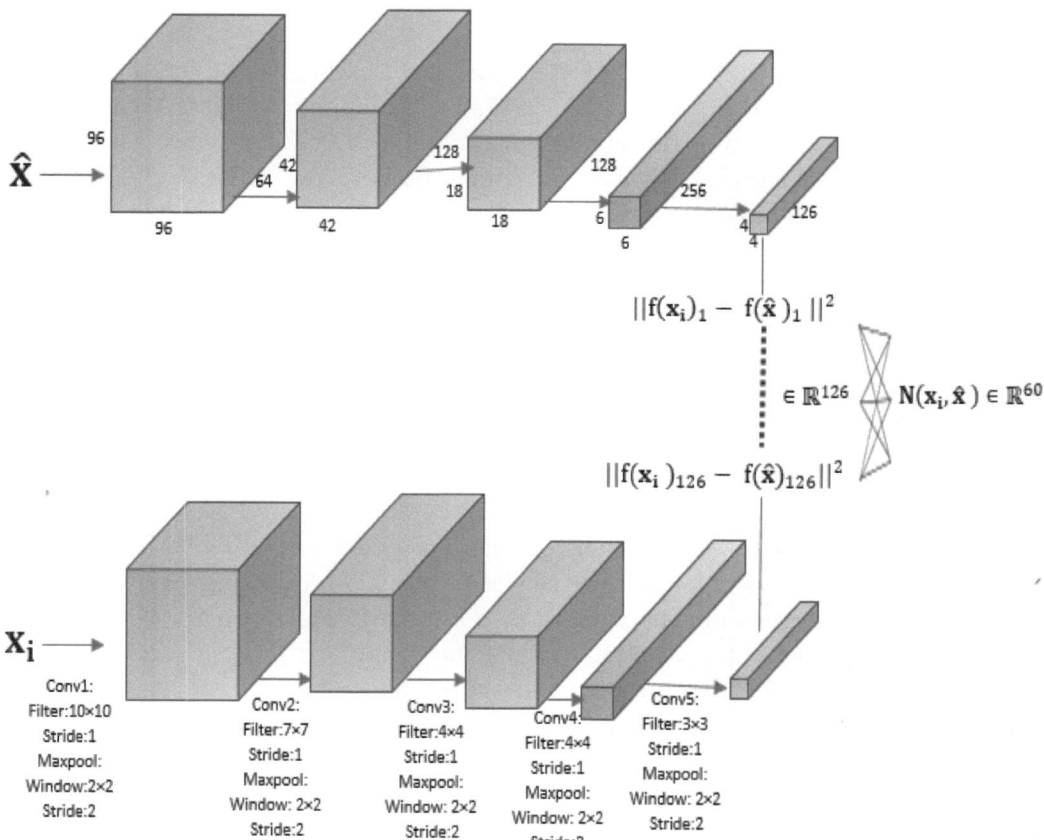

Figure 9.1 Proposed network taking input sample $\hat{\mathbf{x}}$ from the query set and $\mathbf{x_i}$. from the support set. $f(\hat{\mathbf{x}})_j$ represents the j-th feature map of the final convolutional layer embeddings of input $\hat{\mathbf{x}}$. The weighted combination of the squared Euclidean distance between the feature maps of the final convolutional layer is passed through a sigmoid activation function. The negative summation of the output of $N(\mathbf{x_i}, \hat{\mathbf{x}})$ is the measure of similarity between $\hat{\mathbf{x}}$ and $\mathbf{x_i}$.

Where c indicates the level of similarity between the neural network representation of two samples. There have been a variety of similarity functions observed in the literature. Vinyals et al., 2016 and Sachin Ravi and Hugo Larochelle, 2017 applied cosine distance in Matching Networks. Prototypical net (Snell et al., 2017) has achieved a better performance using squared Euclidean distance between the neural networks embedding of samples belonging to query and support set. In our approach, we have used two similarity functions throughout the training process.

$$C(f(\mathbf{x_i}), f(\hat{\mathbf{x}})) = -\sum_{j=1}^{126} ||f(\mathbf{x_i})_j - f(\hat{\mathbf{x}})_j||^2 \qquad (9.1)$$

$$C(f(\mathbf{x_i}), f(\hat{\mathbf{x}})) = -\sum_{j=1}^{60} N(\mathbf{x_i}, \hat{\mathbf{x}})_j \qquad (9.2)$$

It is worth noting that the similarity function 1 is actually the same as squared Euclidean distance used in Snell et al., 2017. For similarity function 2, we consider the weighted combination of the squared distances between the feature maps passed through a sigmoid activation function determines the level of similarity of a particular attribute between two samples. For $N(\mathbf{x_i}, \hat{\mathbf{x}})_j = 0$, the j-th attribute of sample $\mathbf{x_i}$ and $\hat{\mathbf{x}}$ are identical. If $N(\mathbf{x_i}, \hat{\mathbf{x}})_j = 1$, the attribute is different. We have arbitrarily assumed that there could be 60 attributes of a sample and the overall similarity between any two images is the negative summation of these similarity levels. Given a query $\hat{\mathbf{x}}$ and its label encoded as one hot representation $\mathbf{t}$, if the predicted label is $\hat{\mathbf{y}}$, the loss function for the query can be calculated as,

$$\text{Loss}(\hat{\mathbf{y}} \mid (\hat{\mathbf{x}}, \mathbf{t})) = -\log(\text{P}(\hat{\mathbf{y}} \mid (\hat{\mathbf{x}}, \mathbf{t}))) = -log(\prod_{i=1}^{k} \hat{y}_i^{t_i}) \qquad (9.3)$$

Here $\hat{y}_i$, t_i represents i-th element of vector $\hat{\mathbf{y}}$ and $\mathbf{t}$ respectively. The one-shot learning process consists of minimizing the loss function for each query $(\hat{\mathbf{x}}, \mathbf{t}) \in Q$. For the training of a deep neural network with the Omniglot dataset, k random classes from the background set are selected, then one sample from each class is chosen for the support set. And from the rest of the samples of each class, q samples are randomly chosen to form the query set. During the first 400k episodes, we have used similarity function 1 and set k=20, q=1, learning rate=2.2e-05. Here, the 20 classes are randomly chosen regardless of the alphabets from the 964 background classes.

In the last 40k episodes when similarity function 2 was used, we set k=30, q=5, learning rate= 5e-06, with uniform decay of .9999 per episode. Here, the episodes alternate between choosing k classes regardless of the alphabets and choosing them from within a randomly selected alphabet. For those background alphabets whose number of classes are less than 30, the remaining classes are selected from another randomly chosen alphabet. Following the same strategy, we also selected the query and support sets from the validation set, which we used to keep track of the loss during the training process.

9.4 EVALUATION ON OMNIGLOT

The parameters of the deep neural network model were initialized with Glorot uniform. Unlike (Koch et al., 2015), we have not used the validation set to determine

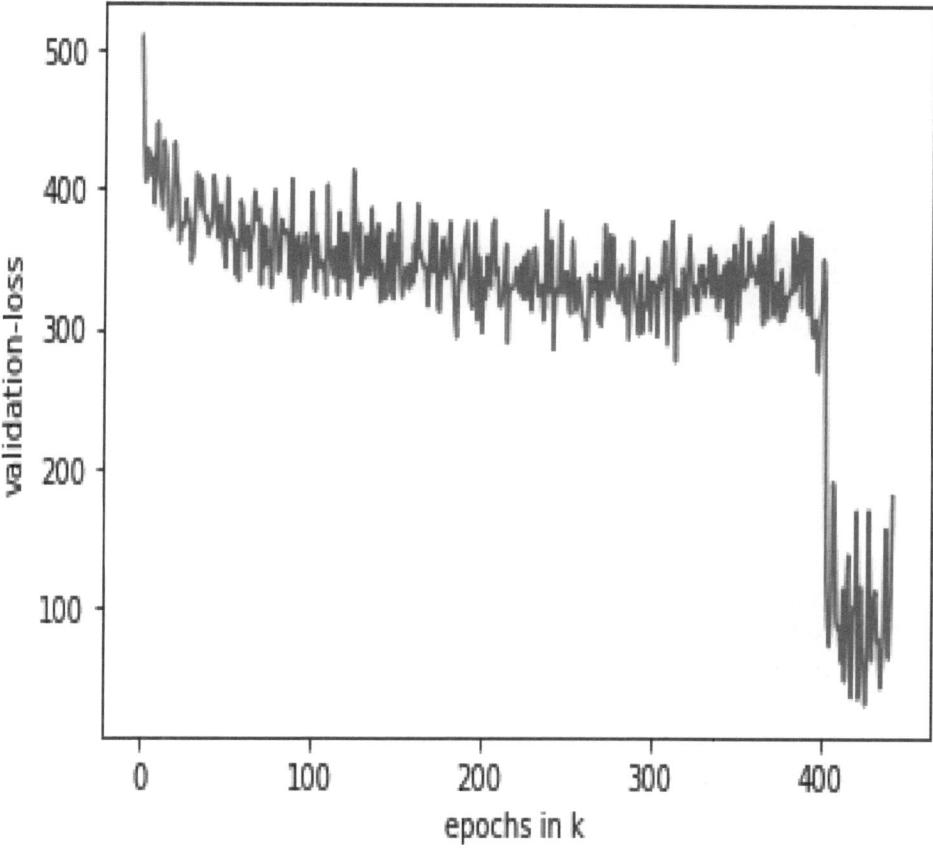

Figure 9.2   Validation loss during the training process

when to stop the training process. Rather, we have tested the model at 440k episodes of training. For each of the 400 queries belonging the original Omniglot challenge (Lake, et al., 2019) and their corresponding support set, we calculated $\hat{\mathbf{x}}$ using similarity function 2. The classification of any query $(\mathbf{x}, \mathbf{t})$ is correct if $argmax_i y_i$ is equal to $argmax_i t_i$. After repeating the training and the evaluation process for 20 times, we took the average classification accuracy as our final result. We also calculated the macro precision and recall for each of the 20 one-shot learning settings in Omniglot challenge and their average turned out to be .95 and .925, respectively.

9.5   EVALUATION ON MNIST

The query set for testing the one-shot learning model of this chapter on MNIST dataset can be represented as $\{(\{\mathbf{x_{ij}}\}_{j=1}^{1000}, \mathbf{y_i})\}_{i=0}^{9}$, where $\mathbf{x_{ij}}$ is the j-th sample of label i in the MNIST test set, whereas the support set can be given as $\{(\mathbf{x_i}, \mathbf{y_i})\}_{i=0}^{9}$. Here, $\mathbf{x_i}$ is a randomly chosen sample of label i form the MNIST training set. Before carrying out the evaluation on the trained model, we took the MNIST dataset in

Table 9.1 Performance of different algorithms on the original Omniglot challenge. Each model presented here has learnt from the Omniglot background set without any data augmentation.

Models	Omniglot accuracy
Bayesian Program Learning	96.7
Humans (Lake et al., 2015)	>95.5
Proposed Net	95.75
Hierarchal Bayesian Program Learning	95.2
Recursive Cortical network	92.75
Simple ConvNet (Lake et al., 2015)	86.5
Prototypical net	86.3
Variational Homoencoder (Hewitt et al., 2018)	81.3

binary format and inverted the samples. Then we upsampled them by 3.75 times to make MNIST dataset compatible with the Omniglot samples. The model is trained in similar fashion with the Omniglot background set, except that in the last 40k episodes, we set k=10 and the classes were chosen regardless of the alphabets. Here, we also repeat the training and evaluation process for 20 times. The average classification accuracy thus obtained is compared with similar experiments found in the literature in table 9.2.

Table 9.2 One-Shot Classification accuracy on MNIST dataset

Models	MNIST accuracy
Proposed Net	78.3
Recursive Cortical network	76.6
Matching net	72
Convolutional Siamese net (Koch et al., 2015)	70.3

9.6 CONCLUSION

In this chapter, we have introduced a measure of similarity for training a matching network to perform one-shot classification. The training process did not involve any class or sample augmentation. Also, we have evaluated our model on the original within-alphabet Omniglot challenge. The superior performance of our model over using other distance metrics is due to the fact that making use of the similarity function makes it capable of choosing which feature difference to ignore while emphasizing feature differences that are more relevant in defining the similarity of an arbitrary attribute between two images. We also believe that the attributes represented by the weighted combination of a trained model will be specific to the query and support samples from any unknown distribution. Also, the similarity function introduced in this chapter helps in stabilizing the model performance, which is why we did not need to use the validation set to intervene in the training process. However, for the similarity function to work, we have to train the model with the squared Euclidean distance metric first, which is one disadvantage of our study. Moreover, non-iterative (Wang et al., 2018) training is not suitable for our approach since, in order for the model to be dataset agnostic, it is necessary that it is trained with a large number of episodes consisting of query and support sets formed by both within-alphabet and between-alphabet samples.

Bibliography

[Koch et al., 2015] G Koch, R Zemel, and R Salakhutdinov. Siamese neural networks for one-shot image recognition. In ICML Deep Learning Workshop, 2015.

[Vinyals et al., 2016] Oriol Vinyals, Charles Blundell, Tim Lillicrap, Daan Wierstra, et al. Matching networks for one shot learning. In Advances in Neural Information Processing Systems, pages 3630–3638, 2016.

[Snell et al., 2017] Snell, J., Swersky, K., and Zemel, R. S. (2017). Prototypical networks for few-shot learning. In Advances in Neural Information Processing Systems.

[J. Speth and E. M. Hand, 2019] J. Speth and E. M. Hand, "Automated label noise identification for facial attribute recognition," in Proc. IEEE Int. Conference on Computer Vision Pattern Recognition Workshops (CVPRW), Jun. 2019, pp. 25–28.

[George et al., 2017] George, D., Lehrach, W., Kansky, K., Laan, C., Marthi, B., Lou, X., Meng, Z., Liu, Y., Wang, H., Lavin, A., and Phoenix, D. S. (2017). A generative vision model that trains with high data efficiency and breaks text-based CAPTCHAs. Science, 2612(October):1–18.

[X. Liu et al., 2019] X. Liu, F. Zhou, J. Liu, and L. Jiang, "Meta-learning based prototype-relation network for few-shot classification," Neurocomputing in Science-direct, vol. 383, pp. 224–234, Mar. 2020.

[Lake, et al., 2013] Lake, Brenden M, Salakhutdinov, Ruslan R, and Tenenbaum, Josh. One-shot learning by inverting a compositional causal process. In Advances in neural information processing systems, pp. 2526–2534, 2013

[Lake et al., 2015] Lake, B. M., Salakhutdinov, R., and Tenenbaum, J. B. (2015). Human-level concept learning through probabilistic program induction. Science, 350(6266):1332–1338.

[Sachin Ravi and Hugo Larochelle, 2017] Sachin Ravi and Hugo Larochelle. Optimization as a model for few-shot learning. International Conference on Learning Representations, 2017.

[Lake, et al., 2019] Lake, B. M., Salakhutdinov, R., & Tenenbaum, J. B. (2019). The Omniglot challenge: A 3-year progress report. Current Opinion in Behavioral Sciences, 29, 97–104.

[Hewitt et al., 2018)] Hewitt, L. B., Nye, M. I., Gane, A., Jaakkola, T., and Tenenbaum, J. B. (2018). The Variational Homoencoder: Learning to learn high capacity generative models from few examples. In Uncertainty in Artificial Intelligence.

[Wang et al., 2018] Wang, X., Cao, W. Non-iterative approaches in training feedforward neural networks and their applications. Soft Compute 22, 3473–3476 (2018). https://doi.org/10.1007/s00500-018-3203-0.

CHAPTER 10

A Real-Time Classification Model for Bengali Character Recognition in Air-Writing

Mohammed Abdul Kader

Dept. of Electrical and Electronic Engineering, International Islamic University Chittagong, Chattogram,Bangladesh

Muhammad Ahsan Ullah

Dept. of Electrical and Electronic Engineering, Chittagong University of Engineering and Technology, Chattogram, Bangladesh

Md Saiful Islam

Dept. of Electronics and Telecommunication Engineering, Chittagong University of Engineering and Technology, Chattogram, Bangladesh

CONTENTS

I N air writing, hand gestures in three-dimensional space are converted into visual linguistic characters. This method has already established itself as a natural, convenient, and intuitive way of human-computer interaction (HCI). Numerous studies on air writing have been performed, and different methods have been proposed. However, the majority of studies are based on the air-writing method of English character identification. To the best of our knowledge, relatively few studies have been conducted on air-writing for Bengali characters. This research represents a hand gesture-based real-time classification model for Bengali character recognition in air-writing. A data acquisition tool is developed to collect real-time information about hand movement. This device reads the orientation and acceleration of hand gestures in three-dimensional space when a user writes a character in the air. Using Bluetooth, these data are transmitted in real-time to a computer. A data set is prepared taking the real-time air writing data of Bengali characters. This data set is used to train a K-NN classifier. The accuracy of the classification model is 96.7%. Finally, the trained classification model is used in real-time air-writing recognition of Bengali characters.

10.1 INTRODUCTION

Presently we are accustomed to the use of touch screens to interact with the digital world both for input and output purposes. It is predicted that the next-generation technology will connect us with the digital world without the existing physical devices such as smartphones, computers, etc. Instead, the next generation of technology is likely to disappear into us, as extensions of our cognition, and would connect us seamlessly with the digital world '[Tomer Yanay & Erez Shmueli (2020)]'. The development of virtual and augmented reality is anticipated to lead to the replacement of the current display components with dedicated eyeglasses that project the output directly into the users' eyes in the coming days. However, the input scheme for the next generation of technology is not converged yet into a single favorable approach. Speech and gesture are two highly studied approaches that are thought to be the way of inputting human instructions into the digital world. However, both approaches have some major limitations. In a noisy environment, speech is not an efficient way to interact with devices. Additionally, this approach is unsuitable in social places since users cannot connect with the digital world while keeping privacy. The second input method is gesture recognition, which came into focus in recent years. With this technique, users can control digital devices using a small number of predefined gestures, such as distinct, unique hand movements. As the set of gestures is limited, all available interaction options might not be addressed by this input method. To increase the interaction options, we can think about adding more gestures in the gesture set. However, it will not be a natural and user-friendly solution as the users have to learn and memorize all the unique movements that include the gesture set. To overcome the limitation of gesture recognition, the concept of air writing is introduced. In air writing, users should not learn any unfamiliar or unusual set of movements; instead, they should move their hand in the air to write the linguistic character in the air. Therefore, this method might be a single leading inputting approach for the next generation of technology.

Researchers from all across the world have already proposed a significant amount of research on air-writing. The proposed studies can be divided into two categories: writing recognition based on motion sensors and writing recognition based on computer vision. In computer vision-based air-writing recognition, a time series of images is taken when the user moves a finger or any fixed color object to write a character in the air. Then computer vision algorithms are applied to these images to detect, segment, and eventually recognize the gesture. Some computer vision-based air-writing research works are listed in '[Md. S. Alam et al.(2020)]', '[O. De et al.(2016)]', '[S. Poularakis& I. Katsavounidis (2016)]', '[Qu, Chengzhang. (2015)]', and '[Kumar, P. et al.(2017)]'. Our study of these research works finds that most of the proposed models have a high degree of accuracy under a fixed illumination

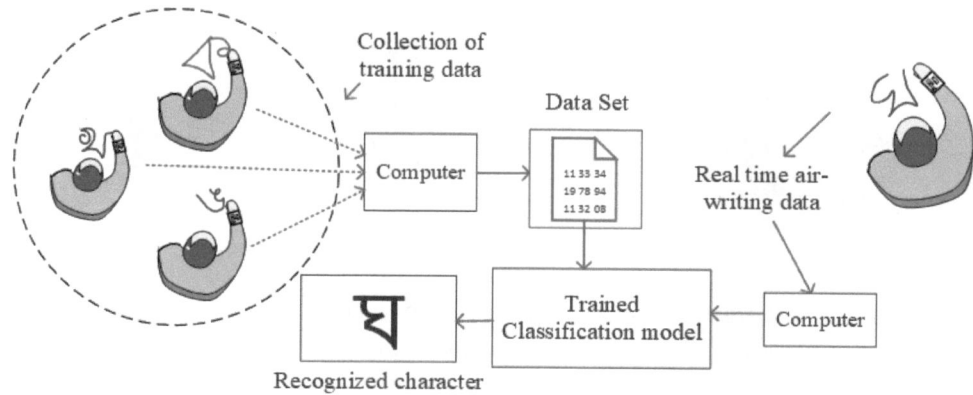

Figure 10.1 Functional diagram of the model.

condition. However, variable illumination and background adversely affect the accuracy of the models. However, the accuracy of gesture-based air-writing recognition is independent of any external factors. In such a method, the user makes a gesture in the air to write a linguistic character while holding or wearing a motion sensor (usually a gyroscope and/or accelerometer). The sensor's data are further analyzed in order to identify the character that the user has written in the air. Numerous algorithms are offered to analyze the motion sensor's raw data and recognize language characters that are listed in '[A. Dash et al.(2017)]', '[M. K. Chakravarthiet al.(2015)]', '[S. Vikram, L. Li, & S. Russell(2013)]','[Yafeng Yin et al.(2019)]', '[Songbin Xu & Yang Xue (2016)]' and '[J. Wang & F. Chuang(2012)]'. However, those research works focused mainly on English character recognition by air-writing.

This chapter represents a hand gesture-based real-time air-writing model for Bengali character recognition. In order to get real-time information about hand movement, a data acquisition device is developed. This device reads the orientation and acceleration of hand movement in three-dimensional space when the user writes a character in the air. These data are transmitted to a computer via Bluetooth. A data set is prepared to take the real-time air writing data of Bengali characters ('ka' to 'ngo'). This data set is used to train a K-NN classifier. The trained classifier is then used to recognize the Bengali character from real-time raw sensor data. The complete block diagram of the model is shown in Figure 10.1.

10.2 METHODOLOGY

In this section, the methodology of the proposed air-writing model is discussed. The three main stages of developing the proposed model are data acquisition, feature extraction, and classifier selection.

10.2.1 Data Acquisition

To collect real-time hand gesture data, a 3-axis accelerometer (ADXL345) is used in this experiment. ADXL345 is an ultra-low power 3-axis digital accelerometer and it

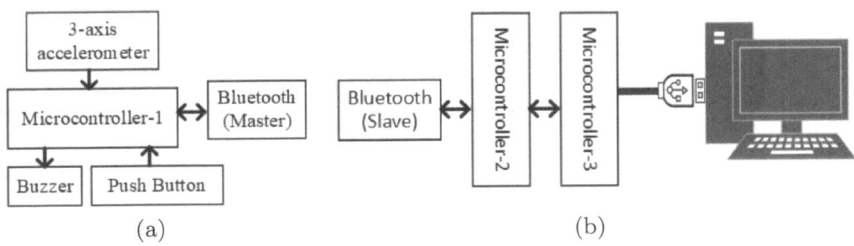

Figure 10.2 Block diagram of Data acquisition circuitry. Illustration in (a) Data acquisition transmitter circuit, in (b) Data acquisition receiver circuit

is accessible through I2C or SPI (3- or 4-wire) digital interface. A microcontroller (ARDUINO Nano development board) is used to read data from ADXL345. The I2C pins (SCL and SDA) of microcontroller are connected to the SCL and SDA pins of ADXL345, respectively. With a sampling rate of 128 Hz, the microcontroller reads the x, y, and z values from the accelerometer sensor. Microcontroller does not read sensor when in idle condition.A push button and a piezo buzzer are attached to the microcontroller's pins D11 and D12, respectively. The microcontroller holds for a second after a user presses this pin, and the buzzer makes a beef sound. The sound indicates that the sensor data has started to be read by the microcontroller. When the user hears the beef sound, they have to immediately begin air-writing. Air-writing have to complete within two seconds. Microcontroller plays another beef sound after two seconds, indicating that it has stopped reading the sensor. It reads the sensor's 768-byte (256-byte data for each axis) data in only two seconds. In the meantime, microcontroller transmits the data via Bluetooth to another circuit, which is connected to computer through USB. The type of communication between two Bluetooth devices is asynchronous. That's why the data bytes obtained from the accelerometer sensor are combined into a single string before transmitting it to the receiver Bluetooth device. Otherwise, after receiving the data, the receiver will fail to distinguish the x-value, y-value and z-value. The data is received as a string by the Bluetooth device at the receiving end, and it is transferred to the microcontroller. The microcontroller converts the string type data into number and separates the x-value, y-value, and z-value from the combined data. Finally, the computer reads the gesture data from this microcontroller circuit. The block diagram of the data acquisition circuitry is shown in Figure 10.2. A sample plot of data for the air-writing of Bengali character 'ka' is shown in Figure 10.3.

10.2.2 Feature Extraction

In general, the data obtained from the air-writing of a character can be represented as follows

$$x = \{x_0, x_1, \ldots, x_{255}\}; \quad y = \{y_0, y_1, \ldots, y_{255}\}; \quad z = \{z_0, z_1, \ldots, z_{255}\} \quad (10.1)$$

The classification model has not been trained directly by the raw data obtained

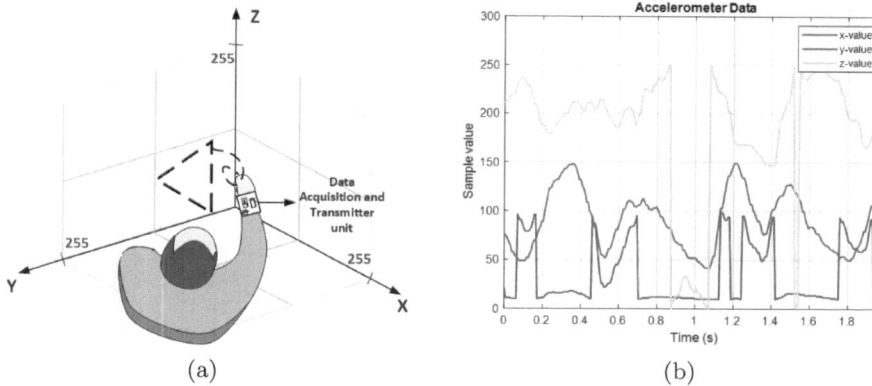

(a)   (b)

Figure 10.3 Air writing in three-dimentional space. Illustration in (a) air-writing of Bengali character 'ka' in three-dimentional space, in (b) Graphical representation of data obtained for air-writing of Bengali character 'ka'

from the sensor. To increase the training and real-time classification speed, the dimentionality of raw data needs to be reduced by removing redundant data. To eliminate data redundancy, three features are extracted from raw sensor data. The features are the mean, standard deviation and first principle component. The mean and standard deviation of x, y and z are calculated by Equation 10.2 and Equation 10.3 respectively.

$$\bar{x} = \frac{\sum_{i=0}^{N-1} x_i}{N}; \quad \bar{y} = \frac{\sum_{i=0}^{N-1} y_i}{N}; \quad \bar{z} = \frac{\sum_{i=0}^{N-1} z_i}{N} \tag{10.2}$$

$$SDx = \sqrt{\frac{\sum_{i=0}^{N-1}(x - \bar{x})^2}{N}}; \quad SDy = \sqrt{\frac{\sum_{i=0}^{N-1}(y - \bar{y})^2}{N}}; \quad SDz = \sqrt{\frac{\sum_{i=0}^{N-1}(z - \bar{z})^2}{N}} \tag{10.3}$$

Principle component analysis (PCA) is performed to get the principle components of data. The PCA process involves the normalization of the data, computation of the covariance matrix, and identification of the eigenvalues and eigenvectors. Finally, principle components are found using the eigen vector of the covariance matrix. Standardization of data can be done by subtracting the mean from the data sample and dividing it by the standard deviation for each value as represented in Equation 10.4.

$$x_i = \frac{x_i - \bar{x}}{SD_x}; \quad y_i = \frac{y_i - \bar{y}}{SD_y}; \quad z_i = \frac{z_i - \bar{z}}{SD_z}; \tag{10.4}$$

Equation 10.5 and Equation 10.6 are used to compute the covariance and covariance matrix respectively.

$$cov(x, y) = \sum_{i=0}^{N-1} \frac{(x_i - \bar{x})(y_i - \bar{y})}{N - 1} \tag{10.5}$$

$$S = \begin{bmatrix} Cov(x, x) & Cov(x, y) & Cov(x, z) \\ Cov(y, x) & Cov(y, y) & Cov(y, z) \\ Cov(z, x) & Cov(z, y) & Cov(z, z) \end{bmatrix} \tag{10.6}$$

The eigen value of covariance matrix 'λ' is obtained by Equation 10.7. The eigen vector of first principle component 'U_1' is calculated by Equation 10.8 taking the maximum eigen value obtained from Equation 10.7.

$$det(S - \lambda I) = 0 \tag{10.7}$$

$$(S - \lambda_{max}I)U_1 = 0 \tag{10.8}$$

Finally, the normalized eigen vector and first principle components are computed using Equation 10.9 and Equation 10.10 respectively.

$$e_1 = \begin{bmatrix} \frac{U_1}{\sqrt{U_1^2+U_2^2+U_3^2}} \\ \frac{U_2}{\sqrt{U_1^2+U_2^2+U_3^2}} \\ \frac{U_3}{\sqrt{U_1^2+U_2^2+U_3^2}} \end{bmatrix} \tag{10.9}$$

$$P_{1i} = e_1^T \begin{bmatrix} x_i - \bar{x} \\ y_i - \bar{y} \\ z_i - \bar{z} \end{bmatrix} \tag{10.10}$$

10.2.3 Classification model

One of the most simple and versatile classification algorithms in machine learning is the k-nearest algorithm (K-NN). In the proposed system, this classification algorithm has been used to recognize the air-writing characters. The main intuition behind the development of the K-NN algorithm is that similar things exist in close proximity. The distances of the new data point from the labeled or trained data points are calculated. The distance may be Euclidean or Cosine or Minkowski distance. Then K-NN finds the k number of nearest data points from the new data point and counts the number of data points for each class. The label or class which is voted maximum by the k number of neighbor data points is predicted as a class or label of the new data point. In this experiment, an application named 'classification learner' has been used to train and validate the classification model. Different classifier models are included in this classification learner app (CLA). For K-NN classifier, the CLA includes Fine K-NN, Medium K-NN, Coarse K-NN, Cosine K-NN, Cubic K-NN and weighted K-NN. The flow chart given in Figure 10.4 shows a common workflow for training classification models in the CLA.

10.3 RESULT AND ANALYSIS

In this chapter, the air-writing model is trained by air-writing data set of five Bengali characters (five consonants: ka ক, kha খ, ga গ, gha ঘ, ngo ঙ). For each character, 60 air-writing samples have been taken. Therefore, total number of air-writing samples for five characters is 300. The duration of each air-writing is 2 sec with a sampling interval of 7.8125 ms. After extracting the features from the raw data set, the KNN classifiers have been trained using the classification learner app (CLA).Cross-validation,

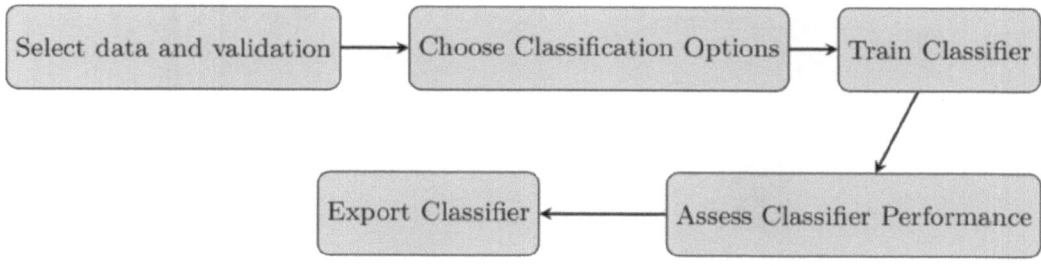

Figure 10.4 Workflow for training classification models in CLA

Table 10.1 Parameter and Performance of different K-NN Classifier

Classifier Name	Number of Neighbour	Prediction Speed (Obs/sec)	Training Time	Distance Metrics	Accuracy
Fine KNN	1	1200	3.2006	Euclidean	93.3%
Medium KNN	10	970	0.80167	Euclidean	93.3%
Coarse KNN	100	5200	0.78231	Euclidean	80.0%
Cosine KNN	10	4900	0.0.92021	Cosine	93.3%
Cubic KNN	10	3800	0.81202	Minkowski	93.3%
Weighted KNN	10	7600	0.76735	Euclidean	**96.7%**

holdout validation, and no validation are the three types of validation available for the CLA. Validation measures the performance of a model on new data and helps the user to choose the best model. The option 'holdout validation' in CLA has been chosen for this work. The CLA divides the dataset into two parts for holdout validation: the training set and the test set. The models are then trained using test data, and the test data are used to estimate the model's performance. In CLA, 25% of the data are defined as the test set and 75% as the training set. The neighbors that each classifier is assuming, together with the performance of the classifiers when the training session is through, are listed in Table 10.1. Considering the accuracy, prediction speed, and training time of the classifiers, the most efficient algorithm for the data set is the weighted KNN. The accuracy of the weighted KNN classifier is 96.7% and the prediction speed is 7600 obs/sec, which is the highest among other classifiers. For this reason, for real-time classification of air-writing data, this weighted KNN classifier is used.

The confusion matrix is shown in Figure 10.5(a) for the weighted KNN classifier. The confusion matrix shows that the percentage of correct predictions, or classification accuracy for the character ka (ক), kha(খ), ga(গ), and ngo(ঙ) is 100%. However, classification accuracy of weighted KNN for the character gha(ঘ) is 83.3%. The model misclassified gha(ঘ) as ga(গ) in 16.7% cases. The relative operating characteristic curve (ROC) of the weighted KNN model for the classification of the character 'ga' is shown in Figure 10.5(b). This curve shows the true positive rate (TPR) with respect to the false positive rate (FPR). The area under curve (AUC) of the ROC curve is 1 for a perfect classifier. In this case, the AUC is 0.98, which represents high sensitivity

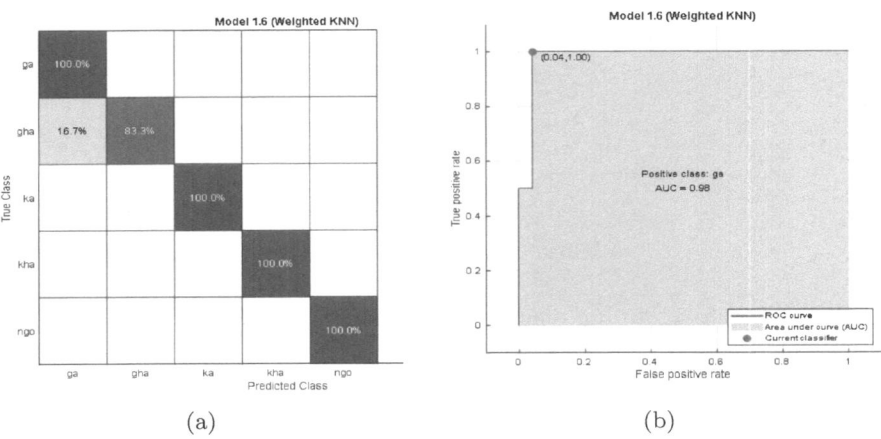

Figure 10.5 Illustration in (a) Confusion matrix of weighted KNN classifier, in (b) ROC curve

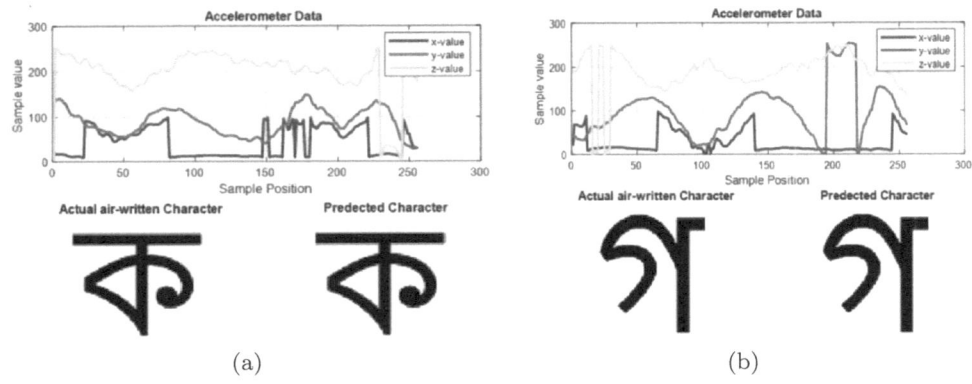

Figure 10.6 Illustration in (a) air-writing recognition of ka(ক) in real-time, in (b) air-writing recognition of ga(গ) in real-time

(very low false negatives) of the model in recognizing the character 'ga'. AUCs for other characters are also found to be very close to 1.

After completing the training session, the most efficient model (weighted KNN) has been exported to the workspace from CLA. Then a program has been developed to evaluate the performance of the model in real-time. The program collects real-time air-writing data from the data acquisition device, computes the features from raw data, and finally classifies the data using the trained weighted KNN classifier. Some outputs of real-time recognition of the proposed model are shown in Figure 10.6 and Figure 10.7. The figures represent four outputs. In the first three outputs, Figure 10.6 (a,b) and Figure 10.7(a), the model successfully classified the character written in the air. The output in Figure 10.7(b) shows that the model failed to classify correctly. The actual air-written character was 'ga', however, the model classified the character as 'gha'.

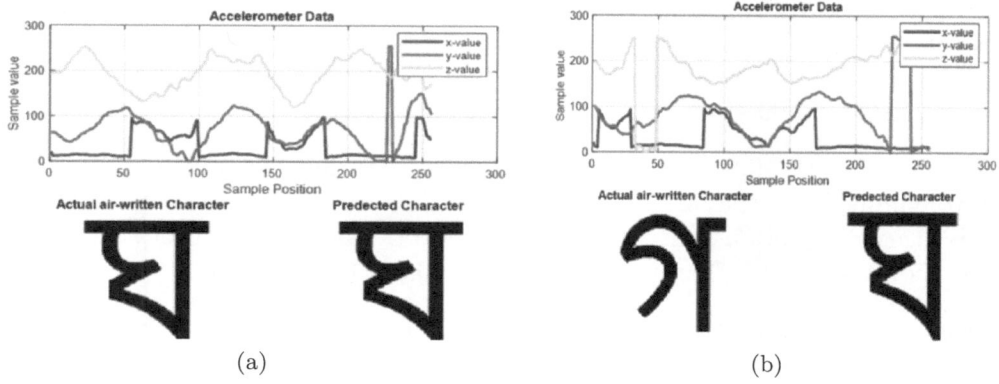

Figure 10.7 Performance of the proposed model in real-time for the characters gha(ঘ) and ga(গ); Illustration in (a) Air-writing recognition of gha(ঘ) in real- time, in (b) air-writing recognition of ga(গ) in real-time

10.4 CONCLUSION AND FUTURE WORK

In this chapter, an approach for real-time air-writing recognition of Bengali characters is suggested. A data acquisition device is developed and a data set is prepared using that device. The proposed system can effectively recognize these characters when air-writing. However, this research, we focused only on the overall functionality of the system. There are many limitations of this research. The data set is prepared only for five Bengali characters. Standards of preparing the data set are not followed. The data set is only tested for KNN classifiers. No analysis is performed for other existing classifiers or algorithms. Also, the analysis involves only three features. There are possibilities to include more impactful features to improve the efficiency and speed of the model. This research has many scopes to extend in the future. The accuracy of the model can be investigated taking gesture data both from accelerometer and gyro sensors for different sampling rates. A complete data set including all Bengali characters and numbers can be prepared maintaining the standards to analyze the functionality of the model. Finally, different classifiers and algorithms can be tested for the data set to get the best model for Bengali air-writing recognition.

ACKNOWLEDGEMENT(S)

We acknowledge and grateful to the anonymous reviewers for their detailed direction to improve the research output.

DISCLOSURE STATEMENT

There are no disclosed conflicts of interest by the authors of this research.

Bibliography

[A. Dash et al.(2017)] A. Dash et al. (2017). AirScript - Creating documents in air. *14th IAPR International Conference on Document Analysis and Recognition (ICDAR)*, pp. 908-913, doi: 10.1109/ICDAR.2017.153.

[Md. S. Alam et al.(2020)] Alam, Md. S.; Kwon, Ki-Chul; Alam, Md. A.; Abbass, Mohammed Y.; Imtiaz, Shariar M.& Kim, Nam. (2020).Trajectory-based air-Writing recognition using deep neural network and depth sensor. *Sensors, 20*, no. 2 (January 9): 376. http://dx.doi.org/10.3390/s20020376.

[J. Wang & F. Chuang(2012)] J. Wang & F. Chuang(2012).An accelerometer-based digital pen with a trajectory recognition algorithm for handwritten digit and gesture recognition. *IEEE Transactions on Industrial Electronics, 59*, no. 7, pp. 2998-3007, doi: 10.1109/TIE.2011.2167895

[Kumar, P. et al.(2017)] Kumar, P., Saini, R., Behera, S.K., & Dogra, D.P. (2017). Real-time recognition of sign language gesture and air-writing using leap motion. *Fifteenth IAPR International Conference on Machine Vision Applications*, pp. 157-160, doi: 10.23919/MVA.2017.7986825.

[M. K. Chakravarthiet al.(2015)] M. K. Chakravarthi, R. K. Tiwari, & S. Handa. (2015).Accelerometer based static gesture recognition and mobile monitoring system using neural networks. *Procedia Computer Science, 70*,pp. 683-687. https://doi.org/10.1016/j.procs.2015.10.105

[O. De et al.(2016)] O. De, P. Deb, S. Mukherjee, S. Nandy, T. Chakraborty & S. Saha (2016). Computer vision based framework for digit recognition by hand gesture analysis *IEEE 7th Annual Information Technology, Electronics and Mobile Communication Conference (IEMCON)*, pp. 1-5, doi: 10.1109/IEMCON.2016.7746361.

[Qu, Chengzhang. (2015)] Qu, Chengzhang. (2015).Online Kinect handwritten digit recognition based on dynamic time warping and support vector machine. *Journal of Information and Computational Science, 12*,pp. 413-422.

[S. Poularakis& I. Katsavounidis (2016)] S. Poularakis& I. Katsavounidis (2016).Low-complexity hand gesture recognition system for continuous streams of digits and letters. *IEEE Transactions on Cybernetics, 46*, no. 9, pp. 294-2108. doi: 10.1109/TCYB.2015.2464195.

[S. Vikram, L. Li, & S. Russell(2013)] S. Vikram, L. Li, & S. Russell (2013).Handwriting and gestures in the air, recognizing on the fly. *Proceedings of the CHI, 13*, pp.1179-1184.

[Songbin Xu & Yang Xue (2016)] Songbin Xu & Yang Xue (2016).Air-writing characters modelling and recognition on modified CHMM. *IEEE International Conference on Systems, Man, and Cybernetics (SMC)*, pp. 001510-001513, doi: 10.1109/SMC.2016.7844452.

[Tomer Yanay & Erez Shmueli (2020)] Tomer Yanay & Erez Shmueli (2020). Air-writing recognition using smart-bands. *Pervasive and Mobile Computing, 66,* 143–168.

[Yafeng Yin et al.(2019)] Yafeng Yin, Lei Xie, Tao Gu, Yijia Lu, & Sanglu Lu (2019).Air contour: Building contour-based model for in-air writing gesture recognition. *ACM Transactions on Sensor Networks, 15,*Issue 4, pp. 1-25. https://doi.org/10.1145/3343855.

CHAPTER 11

A Deep Learning Approach for Covid-19 Detection in Chest X-Rays

SK. Shalauddin Kabir

Department of Computer Science and Engineering, Jashore University of Science and Technology, Jashore, Bangladesh

Mohammad Farhad Bulbul

Department of Computer Science and Engineering, Pohang University of Science and Technology (POSTECH), Pohang, Republic of Korea;
Department of Mathematics, Jashore University of Science and Technology, Jashore, Bangladesh

Fee Faysal Ahmed

Department of Mathematics, Jashore University of Science and Technology, Jashore, Bangladesh

Syed Galib

Department of Computer Science and Engineering, Jashore University of Science and Technology, Jashore, Bangladesh

Hazrat Ali

College of Science and Engineering, Hamad Bin Khalifa University, Qatar Foundation, Doha, Qatar

CONTENTS

THE NOVEL CORONAVIRUS 2019, called SARS-CoV-2 causing COVID-19 has spread swiftly worldwide. An early diagnosis is more important to control its quick spread. Medical imaging mechanics, chest calculated tomography, and X-rays are playing a vital role in the identification and testing of COVID-19 in this present epidemic. Chest X-ray is a cost-effective medical imaging procedure. Infected individuals are increasing rapidly in number and the conventional X-ray oriented test of them is inefficient. So, developing an automated COVID-19 detection process to effectively control the spread is very important. Here, we address the task of detecting COVID-19 automatically by using a popular deep learning model namely the VGG19 model. We use 1300 healthy and 1300 confirmed COVID-19 chest X-ray images. We perform three experiments by freezing different blocks and layers of VGG19. Finally, we use the SVM classifier for detecting COVID-19. In every experiment, we use a five-fold cross-validation method to train and validate the model and finally achieved **98.10%** overall classification accuracy. The results show that our proposed method using the deep learning-based VGG19 model can play a crucial role in diagnosing COVID-19 and be used as a tool for radiologists.

11.1 INTRODUCTION

Towards the end of 2019, the people of the world faced an epidemic of respiratory syndrome called COVID-19. The disease spread in an unprecedented way and quickly turned into a pandemic. Now, COVID-19 is being considered a terrible health hazard all over the world. In December 2019, it was initially identified in Wuhan, China [1, 2]. Coronavirus has a large family; SARS-CoV-2 is one of them that can attack both animals and humans [3] and causes respiratory disease. Other corona viruses recognized previously were SARS-CoV and MERS-CoV in the Middle East. There were almost 8100 and 2500 verified incidents, and death rates of approximately 9.2% and 37.1% were reported, respectively [4, 5].

At the time of this writing **(March 12, 2021)**, around **224** countries are affected seriously by COVID-19. The exact number of affected and dead people worldwide is around **124.1** million and **2.71** million, respectively. Around **98.5** million people have recovered from this epidemic disease, which plunged mankind into a serious condition of dread whose result is yet uncertain [6].

Generally, the COVID-19 virus spreads when normal people come in close contact with infected people. Aerosols and small droplets carry the virus, which easily spreads from an infected person's mouth and nose when he/she talks, breathes out, coughs, sings, or sneezes. The virus might also spread with the aid of contaminated surfaces, though this is no longer believed to be the primary route of transmission [7, 8]. COVID-19 has various types of symptoms from mild to serious and the symptoms take 2-14 days to appear. General symptoms are fever, cough, chills, shortness of breath, tiredness, aches in the body and muscles, headaches, sore throat, diarrhea, and vomiting. In some serious situations, the infection can create multi-organ failure, septic shock, pneumonia, and even death [9].

The tools initially available to doctors and medical experts battling the disease, were inadequate when the COVID-19 outbreak started in China. The treatment of the virus has reached a disappointing stage in some developing countries for the growing requirement of exquisite care units. At the same time, the hospitals and clinics are packed with COVID-19 patients that are getting worse. On the other hand, COVID-19 tests are generally difficult because there are not enough test kits for people. Also, the general public may panic because of their limited knowledge of COVID-19. Besides, it is tough to provide proper health treatment to a large number of people because people are very quickly infected by the virus. Consequently, initial detection and monitoring of COVID-19 are difficult [10]. We need to rely on other diagnostic methods to detect COVID-19 because there are fewer COVID-19 testing tools.

If we want to control the epidemic situation and prevent the virus from spreading, early detection is crucial. Polymerase Chain Reaction, real-time RT-PCR, and RT-LAMP are used as modern analytic tests for COVID-19 detection [11, 12]. COVID-19 chest scan is an example of the unmistakable pneumonia method to reduce the COVID-19 testing time. Through this strategy, doctors and medical specialists will more quickly find isolation and treatment options for patients.

Patients may get a negative outcome from the RT-PCR test when they are infected by the virus and have symptoms too [11]. In this situation, different medical imaging methods, like chest X-rays or chest computed tomography, are used for COVID-19 diagnoses. Even though CT is used as the most effective diagnostic procedure for Covid detection [13], the process has some significant drawbacks, including the relatively higher cost of ionizing radiation, which is significantly $70 \times$ higher than X-rays [14]. For this reason, it is not regularly utilized in COVID-19 detection [15]. Additionally, it is not appropriate for checking the advancement of explicit cases, especially in sick individuals. X-ray is a less delicate methodology to detect COVID-19 when contrasted with 69% revealed pattern sensitivity of CT [16]. X-ray analysis is a less expensive procedure and is additionally accessible in most clinics. On the other hand, X-rays will probably be the technique of imaging method needed to analyze and identify COVID-19 patients. Moreover, these strategies may introduce limitations in specific patients, for example, pregnant women, as they may harm unborn babies [17].

COVID-19 is an extraordinarily important research topic. Many researchers already have experimented with discovering solutions in these times of emergency. Artificial Intelligence-based mechanized CT image analysis apparatuses are available for detecting and observing virus-affected individuals [18]. Shan et al. [19] proposed a deep learning method using chest CT scan for contamination destinations and programmed division of all lungs. Xiaowei developed a screening framework for recognition of the virus by using deep learning and CT images [20]. From the study of changing radio-graphic images from CT images of COVID-19, Wang et al. [21] built a model. The model can reduce the crucial time for finding the infection and removes COVID-19 graphical highlights before pathogenic testing for clinical determination.

In our work, we have proposed a method for the identification of COVID-19 from chest x-rays with the VGG19 model [22]. We can easily collect those image samples from hospitals. For this reason, it is easily possible to use the proposed method to test COVID-19.

The key points of this chapter are summarized as follows:

- We propose a deep learning method by using an effective pre-trained VGG19 model [22] to detect COVID-19.

- The model is evaluated on a large dataset of chest X-ray images (1300 COVID-19 vs. 1300 healthy).

- For better performance, we do not use all the total blocks and layers of VGG19 and perform three experiments in three ways.

- We freeze the convolutional blocks and fully connected layers to reduce time complexity for COVID-19 detection.

- In our work, we freeze convolutional block_1, convolutional block_2 and use fully-connected layer_1 (FC_1), fully-connected layer_2 (FC_2), and fully-connected layer_3 (FC_3) of the VGG19 model. In the first setting, we use the FC_1 and FC_2 layers and freeze the FC_3 layer. Then we use the FC_2 and FC_3 layers and freeze the FC_1 layer. Finally, we use the FC_1 and FC_3 layers and freeze FC_2.

- Finally, we use the Support Vector Machine (SVM) classifier to identify the COVID-19 infection. Also, we show that our proposed method provides a better result to detect COVID-19 in a short time and the model can be potentially useful for medical experts and doctors.

The remaining part of the chapter is presented as follows:

- Section 11.2 describes the literature review.

- The dataset description is covered in Section 11.3.

- Section 11.4 presents the proposed methodology.

- Section 11.5 briefly discusses the exploratory result.

- Finally, the conclusion of the work is covered in Section 11.6.

11.2   LITERATURE REVIEW

In this part, we analyze a few significant studies on COVID-19 diagnosis to show the importance of deep learning.

Recently, deep architecture-oriented detection of the virus was done using two X-ray image databases: the Adrain Rosebrock dataset [23] and the X-ray database of COVID-19 [24], which comprises 123 images of X-ray. This study achieved the highest accuracy of 90% with DenseNet201 and VGG19 models. However, due to the limitation of a small data set, the overall performance suffers. Hemdan [25] proposed a framework known as COVID-Net. Seven different pre-trained models were used in the framework namely; ResNetV2, VGG19, Xception, InceptionV3, MobileNetV2, DenseNet201, and InceptionResNetV2. In this study, the model used 50 X-ray images.

Narin et al. accomplished the classification of normal and COVID-19 cases by using deep learning models for binary classification [26]. In this study, ResNet50 achieved the best accuracy of 96%. Sethy et al. [27] used a Convolution Neural Network and obtained 95.38% accuracy by the model to classify the healthy and the COVID-19 positive cases. A ResNet model was proposed by Zhang et al. for the detection of COVID-19 applying Grand-CAM technology and finally, achieved 95.2% accuracy [28].

The COVID-Net model was proposed by Wang et al. [29] for COVID-19 diagnosis using multi-class datasets (healthy, COVID-19, and pneumonia) and a pre-trained ImageNet model. This study used 13,975 images for the experiment among Resnet50, VGG19, and the developed model and achieved 93.3% accuracy [30, 31, 32, 33].

Ozturk [34] developed a model name DarkNet using deep learning for detecting COVID-19. In this model, two different datasets were used, namely, normal images of chest X-ray dataset [35] and the dataset of COVID-19 positive [24]. Finally, the experiment obtained 95.08% accuracy. Afshar et al. developed a CNN-based COVID-CAPS model and got 95.7% accuracy [36]. For the experiment, two different datasets were used for better performance [37].

Similarly, Oh et al. used a pre-trained ResNet18 model to detect COVID-19 [38]. The model achieved 88.9% accuracy and different datasets were used as the USNLM dataset [39], CoronaHack [40], and JSTR dataset [41, 42, 43].

11.3   DATASET DESCRIPTION

In this section, we describe the collection and preparation of the dataset.

11.3.1   Data collection

For our work, we use chest X-ray images to detect positive or negative COVID-19 infection. The images are taken from different open-source platforms. Here, we use 2600 chest X-ray images including both healthy and COVID-19 positive people. The details are given below:

1. A total 100 of COVID-19 affected X-ray images were collected from Kaggle, which is published by Jocelyn Zhu [44]. Also, 250 healthy people's images were collected from Kaggle's Chest X-Ray dataset [37].

2. Furthermore, 50 images of infected chests were collected from the relevant dataset of Agchung [31].

3. A total of 480 images of Covid positive and 438 images of normal chest X-rays were collected from the BIMCV database [45, 46].

4. Additionally, 290 infectious chest images were obtained from Cohen's dataset [24].

5. A total of 180 images of COVID-affected patients were taken from Agchung [30].

6. Finally, 200 COVID-19++ and 612 normal category chest X-ray images were collected from the database of COVID-19 radiography [33].

11.3.2 Dataset creation

We collected chest X-ray images of healthy and COVID-19 affected people. In the available datasets, these images have some issues, namely incorrect labels, noise, and duplicates. To clean this data, we used the Python NumPy and OpenCV libraries. After cleaning data, we developed our dataset for this work and separated the data into two different classes, i.e., Normal and COVID-19++.

The number of images available as COVID-19++ and Normal in the dataset is shown in Figure 11.1. The sample of Normal and COVID-19++ images in the dataset is shown in Figure 11.2. Table 11.1 indicates the total number of chest X-ray images in the dataset used for training, validation, and testing.

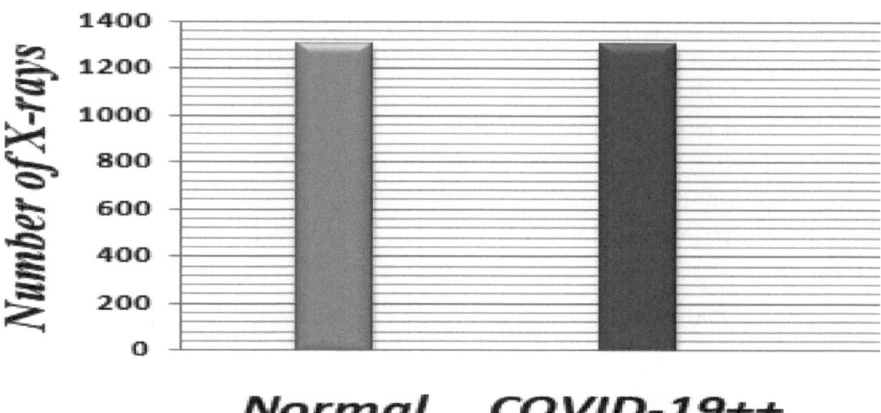

Figure 11.1 Distribution of Normal and COVID-19++ Chest X-ray images in the dataset

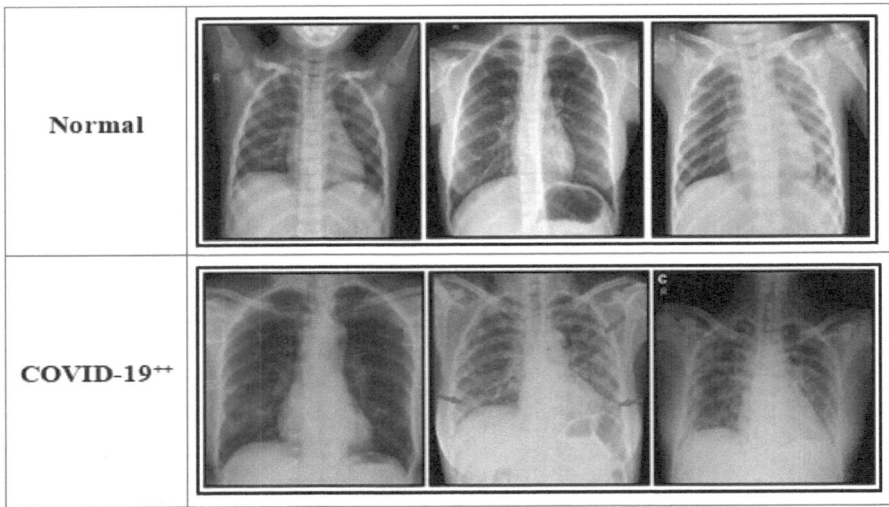

Figure 11.2 The sample of chest X-ray images.

Table 11.1 Chest X-ray samples used in our proposed methodology.

Category	Normal	COVID-19^{++}
Training phase	910	910
Validation phase	130	130
Testing phase	260	260

11.4 PROPOSED METHODOLOGY

At first, we developed an automated COVID-19 detection model using a modified Convolutional Neural Network (CNN) architecture. Then, we trained the model with the features representation mode of transfer-learning. This section provides detailed information about the proposed algorithm, pre-processing, augmentation of image data, the proposed deep neural network, and the transfer-learning model. Also, this section gives detailed information on the necessary settings for the model such as fine-tuning, experimental setup, and model evaluation stages. Figure 11.3 contains the flowchart of our proposed COVID-19 classification methodology.

11.4.1 Proposed Algorithm

Following are the steps in the proposed method:
   Step 1. Image pre-processing (resizing, normalization, and augmentation).
   Step 2. Provide the images as input to the deep learning-based VGG19 model.
   Step 3. Freeze convolutional block_1 and convolutional block_2.
   Step 4. Fetch the output of the last convolutional block.
   Step 5. Apply fully connected layers to the output.
   Step 6. Freeze the fully connected layer.

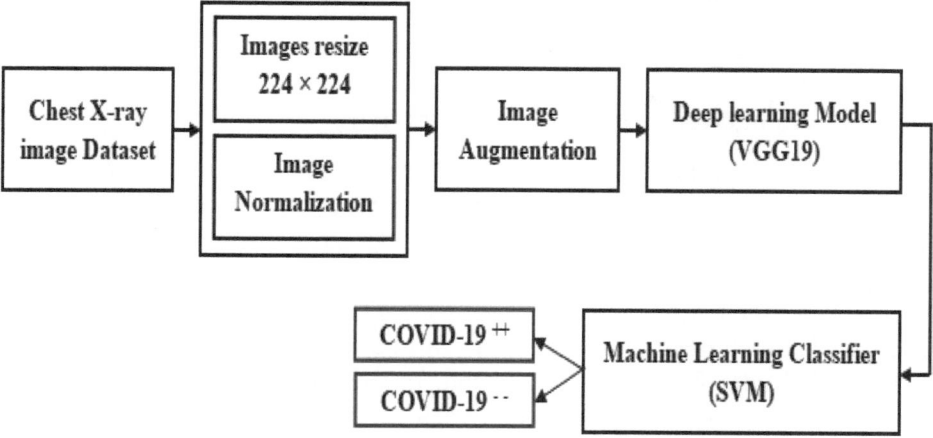

Figure 11.3 Flowchart of our proposed COVID-19 classification methodology.

- First setting: Freeze FC_3 layer.

- Second setting: Freeze FC_1 layer.

- Third setting: Freeze FC_2 layer.

Step 7. Apply Activation map (ReLU).
Step 8. Apply SVM for classification.

11.4.2 Preprocessing: Image resize and normalization

Image resizing and normalization techniques are applied in the pre-processing step to reduce the diversity of the original images.

At first, we convert all the images to gray scale because the images were collected from different sources. In our work, we have used VGG19 as a feature extraction pre-trained model. Since the input size of VGG19 is 224 × 224, every image is resized to 224 × 224 pixels. This helps us reduce the experimental requirements and achieve compatibility of the data with the input layer of the VGG19 model. We can apply the normalization and equalization techniques to correct contrast and brightness. The two techniques provide comparable results. But normalization is simpler than equalization and provides realistic images in the result. Finally, depending on radiation and the mechanism of image acquisition [47, 48], we apply the image normalization process to address the enormous fluctuation of the image outlook (contrast and brightness).

11.4.3 Augmentation of Images

A huge number of data are needed to develop generalized and powerful deep learning-based models. Moreover, clinical imaging information and data are rare, and for this reason, labeling the dataset is costly. To alleviate the problem of model over-fitting, different image augmentation techniques are applied [49] on the dataset such as width shift (a range of 20%), random rotation (to a maximum of 15 degrees), height shift (a range of 20%), and zoom (a range of 20%).

11.4.4 Deep Neural Networks and Transfer-learning

Machine learning has different fields and deep learning technology is one of them. The developed deep learning technique is used to improve the task of image classification. We use Keras, OpenCV, TensorFlow, and Scikit-learn libraries for model training.

Feature extraction and parameter tuning are the two steps for accomplishing the transfer learning technique. In the feature extraction step, the convolutional neural network based pre-trained model uses the training data to extract new features. The VGG19 model is used for feature extraction. In the second step, parameter tuning is used to increase the performance by re-creating and updating the model structure. The pre-trained model used in this study is discussed below.

The CNN-based VGG19 model and SVM machine learning classifier are used to classify COVID-19^{++} or negative from the given dataset. To train the VGG19 model we use the chest X-ray images.

VGG19: In 2014, VGG was first introduced and is also a familiar deep convolutional neural network. VGG has multiple layers that eventually help in improving the model's performance [22]. The main benefit of the VGG model is that it uses only 3×3 convolutional layers. The model has another three fully connected layers. VGG19 is a sub-branch of the family of VGG networks and provides around 143 million parameters for the work of the network. The model has 19 layers. Moreover, the model is used to explore the object of the chest X-ray images. We use the VGG19 model for feature extraction and the SVM machine learning classifier for the detection of COVID-19. SVM can easily perform a nonlinear classification and provides the inputs in high-feature layers. SVM is one of the most used techniques. To replace the Softmax layer in the CNN model, many researchers used the SVM. By changing the layer, they created a new model named CNN-SVM [50, 51]. In this work, we have replaced the Softmax layer in VGG19 with the SVM classifier to predict the COVID-19 Negative/Positive and get the maximum accuracy in our experiment.

11.4.5 Fine-tuning

At first, we remove the common low-level sampling layers by freezing the weights. The first few layers acquire knowledge of generic functionalities that helps to generalize the images. The actual target of fine-tuning is to modify the recently collected chest X-ray dataset with these features.

In this stage, we try to train several layers on the upper level of the base model. It is difficult to improve the pre-trained model's weights during the training period. In this situation, a new network head is attached with the intended categories and prepared for adjusting the weights as indicated by new patterns. We use the fine-tuning technique of the upper levels to increase the performance of the model.

11.4.6 Experimental Setup

In this experiment, we use a pre-trained VGG19 model. We train the model on a Google Colaboratory server. Here, we use the Windows 10 operating system with Intel(R) Core^{TM} i5-4200M CPU @ 2.50GHz. We install TensorFlow 2.0 and Keras (along with the libraries of matplotlib, OpenCV, and scikit-learn) in our working environment. The Adam optimizer is used to train our model with the initialization weights. The rotation_range, batch size, learning rate, and epoch number are set to 15, 12, 1e-3, and 50, respectively for the experiment. A five-fold cross-validation process is used to overcome the model over-fitting problem.

11.4.7 Model Evaluation

The following five criteria are used to calculate the performance of the model:

$$Accuracy = \left(\frac{(TP + TN)}{(TP + TN + FP + FN)} \right) \times 100$$

$$Specificity = \left(\frac{TN}{(TN + FP)} \right) \times 100$$

$$Sensitivity = \left(\frac{TP}{(TP + FN)} \right) \times 100$$

$$Precision = \left(\frac{TP}{(TP + FP)} \right) \times 100$$

$$F1 - Score = \left(2 \times \frac{(Precision \times Sensitivity)}{(Precision + Sensitivity)} \right) \times 100$$

Here, TP means True Positive, TN means True Negative, FP means False Positive, and FN means False Negative.

11.5 RESULTS AND DISCUSSION

In this section, we discuss the model performance and analyze the achieved results.

11.5.1 Evaluation

We perform three experiments to achieve better accuracy in a short time by freezing Convolution block_1 and Convolution block_2 and also freezing fully connected layers. The results from the three experiments are given below.

11.5.1.1 Results on first setting

In our first setting, we freeze the first two convolutional blocks and pooling layers. We also freeze fully-connected layer_3 (FC_3). We use Convolutional block_3, Convolutional block_4, Convolutional block_5 and fully-connected layer_1 (FC_1) and fully-connected layer_2 (FC_2) of VGG19 for completing our experiment. The training phase has been completed up to the 50th epoch to remove the over-fitting problem. We achieve **95.83%** accuracy from our first setting. Table 11.2 contains the performance of the prediction results achieved from the VGG19 model by freezing certain blocks and layers. Figure 11.4 presents the training accuracy and the loss values of the model.

Table 11.2 Accuracy, Specificity, Sensitivity, Precision, F1-score of the first setting are reported.

		Performance Results				
First setting	Model	Accuracy	Specificity	Sensitivity	Precision	F1-score
	VGG19	95.83%	91.67%	100%	97.50%	98.73%

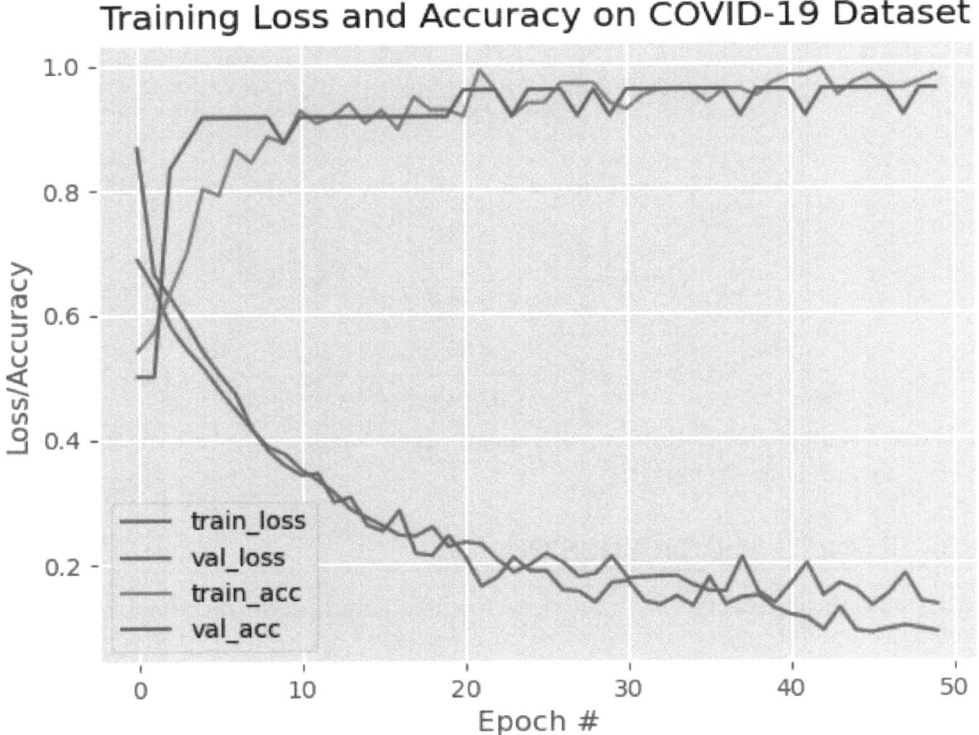

Figure 11.4 Training curves for the training and validation loss and accuracy values for the first setting.

11.5.1.2 *Results on second setting*

In our second setting, we freeze the first two convolutional blocks and pooling layers. We also freeze fully-connected layer_1 (FC_1). We use Convolutional block_3,

Convolutional block_4, Convolutional block_5, and fully-connected layer_2 (FC_2) and fully-connected layer_3 (FC_3) of VGG19 for completing our experiment. The training phase has been completed up to the 50th epoch to remove the over-fitting problem in our Keras and Tensorflow model. We achieve **96.67%** accuracy from our second setting. Table 11.3 contains the performance of the prediction results achieved from the VGG19 model by freezing blocks and layers. Figure 11.5 presents the training accuracy and the loss values of the model.

Table 11.3 Accuracy, Specificity, Sensitivity, Precision, F1-score of the second setting are reported.

Performance Results					
Second setting	Model	Accuracy	Specificity	Sensitivity	Precision	F1-score
	VGG19	96.67%	93.33%	100%	98%	98.99%

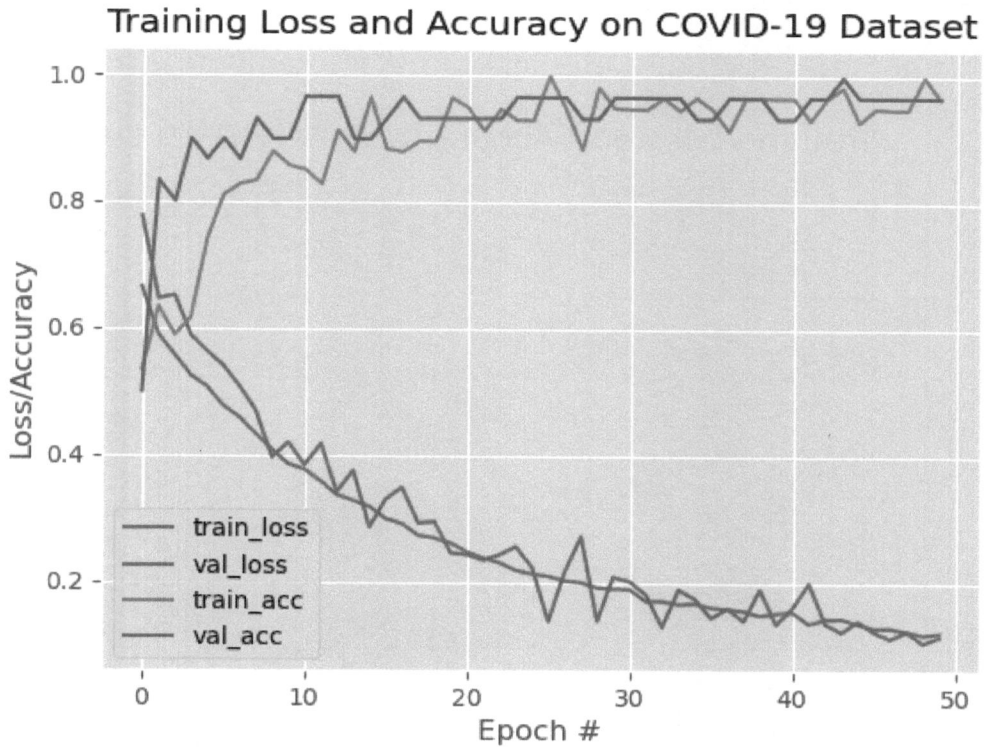

Figure 11.5 Training curves for the training and validation loss and accuracy values for the second setting.

11.5.1.3 *Result on third setting*

In our third setting, we freeze the first two convolutional blocks and pooling layers. We also freeze fully-connected layer_2 (FC_2). We use Convolutional block_3, Convolutional block_4, Convolutional block_5, fully-connected layer_1 (FC_1), and fully-connected layer_3 (FC_3) of VGG19 for completing our experiment. The training phase has been completed up to the 50th epoch to remove the over-fitting problem in our Keras and Tensorflow model. We achieve **98.1%** accuracy from our third setting. Table 11.4 contains the performance of the prediction results achieved from the VGG19 model by freezing blocks and layers. Figure 11.6 presents the training accuracy and the loss values of the model.

Table 11.4 Accuracy, Specificity, Sensitivity, Precision, F1-score of the third setting are reported.

Performance Results					
Third setting	Model	Accuracy	Specificity	Sensitivity	Precision	F1-score
	VGG19	98.10%	95%	100%	98.70%	99.34%

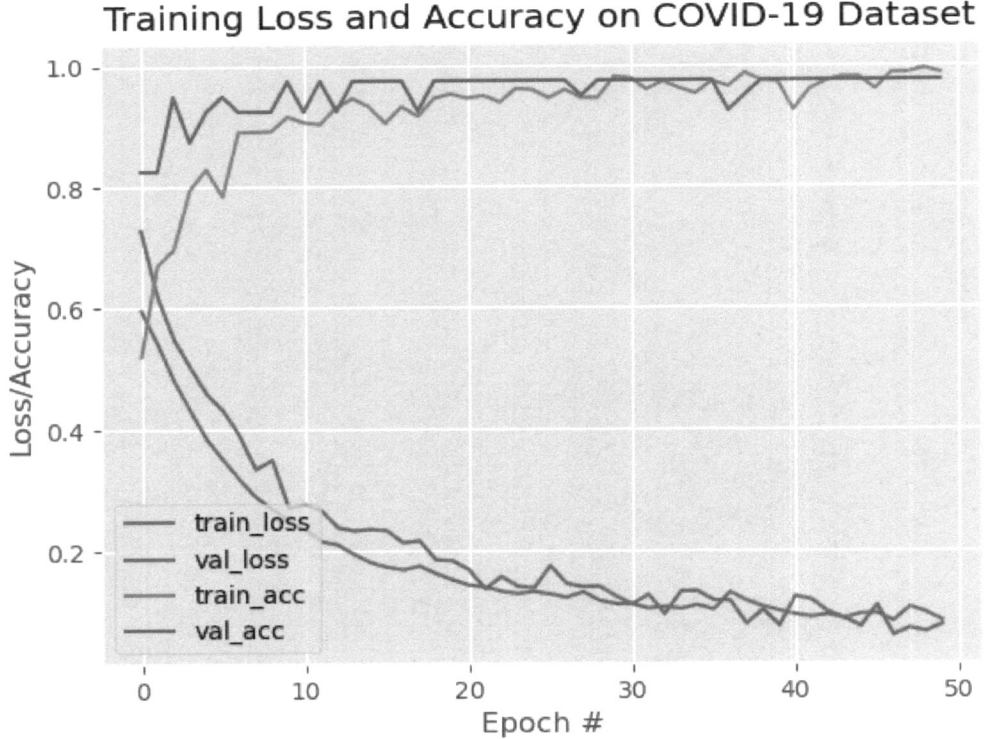

Figure 11.6 Training curves for the training and validation loss and accuracy values for the third setting.

The main advantages of this study are:

1. The experiment is not affected by the imbalanced data.

2. Data augmentation has been used, which helps to improve the overall training of the model.

3. The proposed method reduces the time requirements.

The study has some limitations:

1. For better working performance, we need high-quality equipment.

2. Our proposed method is evaluated only for two classifications, i.e., normal vs. COVID-19 positive classification.

3. Used dataset is not enough for a better result. Need to use the more different dataset for classification.

11.6 CONCLUSION

Early detection and prediction of COVID-19 play an essential role so that patient can minimize financial costs and doctors can reduce the diagnostic time. Deep learning methods are promising for image classification tasks. To protect the spread of the virus, early identification of affected people is crucial. In our chapter, we proposed a model using VGG19 to detect COVID-19 affected people. We performed three experiments and got the highest accuracy in the third setting. The overall performance achieved by the model in the third setting is **98.10%** accuracy, **100%** sensitivity, **95%** specificity, **98.70%** precision, and **99.34%** F1-score. From this study, we believe that it will help the doctor and radiologists make decisions and reduce time in medical practice. In the early stage, to detect COVID-19, this study will help to learn how to use deep transfer learning methods. In the future, through this study, we can easily train the metadata models using the mixed images, if different metadata and medical notes are provided for incubation. These metadata models will help us to predict prognoses and also helps patient management of the pandemic period. In the future, we will use our proposed method to perform different disease classifications like COVID-19^{++} vs. viral pneumonia vs. bacterial pneumonia vs. normal.

Bibliography

[1] Roosa, K., Lee, Y., Luo, R., Kirpich, A., Rothenberg, R., Hyman, J., . . . Chowell, G. b. (2020). Real-time forecasts of the COVID-19 epidemic in China from February 5th to February 24th, 2020. *Infectious Disease Modelling, 5,* 256-263.

[2] Zheng, Y.-Y., Ma, Y.-T., Zhang, J.-Y., & Xie, X. (2020). COVID-19 and the cardiovascular system. *Nature Reviews Cardiology, 17*(5), 259-260.

[3] Cui, J., Li, F., & Shi, Z.-L. (2019). Origin and evolution of pathogenic coronaviruses. *Nature Reviews Microbiology, 17*(3), 181-192.

[4] Momattin, H., Al-Ali, A. Y., & Al-Tawfiq, J. A. (2019). A systematic review of therapeutic agents for the treatment of the Middle East Respiratory Syndrome Coronavirus (MERSCoV). *Travel Medicine and Infectious Disease, 30*, 9-18.

[5] Singhal, T. (2020). A review of coronavirus disease-2019 (COVID-19). *The Indian Journal of Pediatrics, 87*(4), 281-286.

[6] Worldometer. (2021). *COVID-19 Worldwide Statistics.* Retrieved from https://www.worldometers.info/coronavirus/? (Last accessed 12 March 2021)

[7] Huang, C., Wang, Y., Li, X., Ren, L., Zhao, J., Hu, Y., ... others (2020). Clinical features of patients infected with 2019 novel coronavirus in Wuhan, China. *The Lancet, 395*(10223), 497-506.

[8] World Health Organization, et al. (2020). Coronavirus disease (COVID-19): How is it transmitted. *Geneva, Switzerland: WHO.*

[9] Mahase, E. (2020). Coronavirus: Covid-19 has killed more people than SARS and MERS combined, despite lower case fatality rate. *BMJ 368: m641.*

[10] Ai, T., Yang, Z., Hou, H., Zhan, C., Chen, C., Lv, W., ... Xia, L. (2020). Correlation of chest CT and RT-PCR testing for coronavirus disease 2019 (COVID-19) in China: A report of 1014 cases. *Radiology, 296*(2), E32-E40.

[11] Zhai, P., Ding, Y., Wu, X., Long, J., Zhong, Y., & Li, Y. (2020). The epidemiology, diagnosis and treatment of Covid-19. *International Journal of Antimicrobial Agents, 55*(5), 105955.

[12] Zhang, W. (2020). Imaging changes of severe COVID-19 pneumonia in advanced stage. *Intensive Care Medicine, 46*(5), 841-843.

[13] Fang, Y., Zhang, H., Xie, J., Lin, M., Ying, L., Pang, P., & Ji, W. (2020). Sensitivity of chest CT for COVID-19: Comparison to RT-PCR. *Radiology, 296*(2), E115-E117.

[14] Lin, E. C. (2010). Radiation risk from medical imaging. *Mayo Clinic Proceedings, 85*(12), 1142-1146.

[15] Self, W. H., Courtney, D. M., McNaughton, C. D., Wunderink, R. G., & Kline, J. A. (2013). High discordance of chest x-ray and computed tomography for detection of pulmonary opacities in ED patients: Implications for diagnosing pneumonia. *The American Journal of Emergency Medicine, 31*(2), 401-405.

[16] Jacobi, A., Chung, M., Bernheim, A., & Eber, C. (2020). Portable chest x-ray in coronavirus disease-19 (COVID-19): A pictorial review. *Clinical Imaging, 64*, 35-42.

[17] Ratnapalan, S., Bentur, Y., & Koren, G. (2008). Doctor, will that x-ray harm my unborn child? *Cmaj, 179*(12), 1293-1296.

[18] Gozes, O., Frid-Adar, M., Greenspan, H., Browning, P. D., Zhang, H., Ji, W.,... Siegel, E. (2020). Rapid AI development cycle for the coronavirus (COVID-19) pandemic: Initial results for automated detection & patient monitoring using deep learning CT image analysis. arXiv preprint arXiv:2003.05037.

[19] Shan, F., Gao, Y.,Wang, J., Shi, W., Shi, N., Han, M., . . . Shi, Y. (2020). Lung infection quantification of COVID-19 in CT images with deep learning. arXiv preprint arXiv:2003.04655.

[20] Xu, X., Jiang, X., Ma, C., Du, P., Li, X., Lv, S., . . . others (2020). A deep learning system to screen novel coronavirus disease 2019 pneumonia. *Engineering, 6*(10), 1122-1129.

[21] Wang, S., Kang, B., Ma, J., Zeng, X., Xiao, M., Guo, J., . . . others (2021). A deep learning algorithm using CT images to screen for Corona Virus Disease (COVID-19). *European Radiology*, 1-9.

[22] Simonyan, K., & Zisserman, A. (2015, May). Very deep convolutional networks for large-scale image recognition. In Y. Bengio & Y. LeCun (Eds.), *3rd International Conference on Learning Representations, ICLR 2015*. San Diego, CA, USA.

[23] Rosebrock, A. (2020). Detecting COVID-19 in x-ray images with keras, tensorflow, and deep learning. URL: https://www.pyimagesearch.com/2020/03/16/detecting-covid-19-in-x-rayimages-with-keras-tensorflow-and-deep-learning. (Last accessed 16 January 2021)

[24] Cohen, J. P., Morrison, P., & Dao, L. (2020). COVID-19 image data collection. arxiv 2003.11597, 2020. URL: https://github.com/ieee8023/covid-chestxray-dataset.

[25] Hemdan, E. E.-D., Shouman, M. A., & Karar, M. E. (2020). Covidx-net: A framework of deep learning classifiers to diagnose COVID-19 in x-ray images. arXiv preprint arXiv:2003.11055.

[26] Narin, A., Kaya, C., & Pamuk, Z. (2021). Automatic detection of coronavirus disease (COVID- 19) using x-ray images and deep convolutional neural networks. *Pattern Analysis and Applications*, 1-14.

[27] Sethy, P. K., & Behera, S. K. (2020). Detection of coronavirus disease (COVID-19) based on deep features. Preprints2020, 2020030300 (doi: 10.20944/preprints202003.0300.v1).

[28] Zhang, J., Xie, Y., Li, Y., Shen, C., & Xia, Y. (2020). COVID-19 screening on chest x-ray images using deep learning based anomaly detection. arXiv preprint arXiv:2003.12338, 27.

[29] Wang, L., Lin, Z. Q., & Wong, A. (2020). Covid-net: A tailored deep convolutional neural network design for detection of COVID-19 cases from chest x-ray images. *Scientific Reports, 10*(1), 1-12.

[30] Agchung. (2020a). Actualmed-covid-chestxray-dataset. Retrieved from https://github.com/agchung/Actualmed COVID-chestxray-dataset (Last accessed 17 January 2021)

[31] Agchung. (2020b). COVID-19 chest x-ray dataset initiative. Retrieved from https://github.com/agchung/Figure1-COVID-chestxray-dataset (Last accessed 16 January 2021)

[32] North America, R. S. (2020). Rsna pneumonia detection challenge. Retrieved from https://www.kaggle.com/c/rsna-pneumonia-detection-challenge/data (Last accessed 16 January 2021)

[33] Rahman, T. (2020). COVID-19 radiography database. Retrieved from https://www.kaggle.com/tawsifurrahman/covid19-radiography-database (Last accessed 17 January 2021)

[34] Ozturk, T., Talo, M., Yildirim, E. A., Baloglu, U. B., Yildirim, O., & Acharya, U. R. (2020). Automated detection of COVID-19 cases using deep neural networks with x-ray images. *Computers in Biology and Medicine, 121,* 103792.

[35] Wang, X., Peng, Y., Lu, L., Lu, Z., Bagheri, M., & Summers, R. M. (2017). Chestx-ray8: Hospital-scale chest x-ray database and benchmarks on weakly-supervised classification and localization of common thorax diseases. *In Proceedings of the IEEE Conference on Computer Vision and Pattern Recognition* (pp. 2097-2106).

[36] Afshar, P., Heidarian, S., Naderkhani, F., Oikonomou, A., Plataniotis, K. N., & Mohammadi, A. (2020). Covid-caps: A capsule network-based framework for identification of COVID-19 cases from x-ray images. *Pattern Recognition Letters, 138,* 638-643.

[37] Mooney, P. (2019). Chest x-ray images (pneumonia). Retrieved from https://www.kaggle.com/paultimothymooney/chest-xray-pneumonia (Last accessed 17 January 2021)

[38] Oh, Y., Park, S., & Ye, J. C. (2020). Deep learning COVID-19 features on cxr using limited training data sets. *IEEE Transactions on Medical Imaging, 39*(8), 2688-2700.

[39] Jaeger, S., Candemir, S., Antani, S., Wang, Y.-X. J., Lu, P.-X., & Thoma, G. (2014). Two public chest x-ray datasets for computer-aided screening of pulmonary diseases. *Quantitative Imaging in Medicine and Surgery, 4*(6), 475.

[40] Praveen. (2020). CoronaHack-Chest X-Ray Dataset. Retrieved from https://www.kaggle.com/praveengovi/coronahack-chest-xraydataset (Last accessed 17 January 2021)

[41] Shiraishi, J., Katsuragawa, S., Ikezoe, J., Matsumoto, T., Kobayashi, T., Komatsu, K.-i., . . . Doi, K. (2000). Development of a digital image database for

chest radiographs with and without a lung nodule: Receiver operating characteristic analysis of radiologists' detection of pulmonary nodules. *American Journal of Roentgenology, 174*(1), 71-74.

[42] Van Ginneken, B., Stegmann, M. B., & Loog, M. (2006). Segmentation of anatomical structures in chest radiographs using supervised methods: A comparative study on a public database. *Medical Image Analysis, 10*(1), 19-40.

[43] Munawar, F., Azmat, S., Iqbal, T., Grönlund, C., & Ali, H. (2020). Segmentation of lungs in chest X-ray image using generative adversarial networks. *IEEE Access, 8*, 153535- 153545.

[44] Zhu, J. (2020). COVID-19 chest x-rays for lung severity scoring. Retrieved from https://www.kaggle.com/jocelynzhu/covid19-chest-xrays-for-lung-severity-scoring (Last accessed 18 January 2021)

[45] Bustos, A., Pertusa, A., Salinas, J.-M., & de la Iglesia-Vaya, M. (2020). Padchest: A large chest x-ray image dataset with multi-label annotated reports. *Medical Image Analysis, 66,* 101797. Retrieved from https://bimcv.cipf.es/bimcvprojects/padchest/ (Last accessed 18 January 2021)

[46] Vaya, M. d. l. I., Saborit, J. M., Montell, J. A., Pertusa, A., Bustos, A., Cazorla, M., . . . others (2021). BIMCV COVID-19+: A large annotated dataset of RX and CT images from COVID-19 patients. *IEEE Dataport.* Retrieved from https://dx.doi.org/10.21227/w3aw-rv39

[47] Guendel, S., Ghesu, F. C., Grbic, S., Gibson, E., Georgescu, B., Maier, A., & Comaniciu, D. (2019). Multi-task learning for chest x-ray abnormality classification on noisy labels. arXiv preprint arXiv:1905.06362.

[48] Stephen, O., Sain, M., Maduh, U. J., & Jeong, D.-U. (2019). An efficient deep learning approach to pneumonia classification in healthcare. *Journal of Healthcare Engineering, 2019* .

[49] Shorten, C., & Khoshgoftaar, T. M. (2019). A survey on image data augmentation for deep learning. *Journal of Big Data, 6*(1), 1-48.

[50] Agarap, A. F. (2017). An architecture combining convolutional neural network (CNN) and support vector machine (SVM) for image classification. arXiv preprint arXiv:1712.03541.

[51] Tang, Y. (2013). Deep learning using linear support vector machines. arXiv preprint arXiv:1306.0239.

CHAPTER 12

Automatic Image Captioning Using Deep Learning

Toshiba Kamruzzaman

Dept. of ECE, Rajshahi University of Engineering and Technology, Rajshahi, Bangladesh

Abdul Matin

Dept. of ECE, Rajshahi University of Engineering and Technology, Rajshahi, Bangladesh

Tasfia Seuti

Dept. of CSE, Rajshahi University of Engineering and Technology, Rajshahi, Bangladesh

Md. Rakibul Islam

Dept. of ECE, Rajshahi University of Engineering and Technology, Rajshahi, Bangladesh

CONTENTS

Transforming an image to a text or descriptive form has recently gained a lot of research appeal. Creating a sentence with correct semantics and syntactic structure is still a matter of concern. Object recognition, relations between the objects, and different meanings of the same word make this task more difficult. Therefore, an inspection of the attention mechanism has recently achieved great progress. In this chapter, we describe some existing models and then expand these models by integrating BERT, LSTM, and dense models together.

After analysing the results, we found that while training on the same parameters, our new model has shown comparatively less training time than the others and has shown better results for all common metrics (BLEU, METEOR, & CIDEr) on the MS-COCO dataset.

12.1 INTRODUCTION

Image captioning is the technique of interpreting a source image into its corresponding text version. The captioning system helps the audience rapidly grasp the image's concept without having to go over every element. The ultimate goal of an image captioning system is to communicate the picture's primary message in a narrative format. Image captioning is a category of image-to-sequence problem in which the sources are pixels (a digital form of an image). The visual encoding stage, which prepares the input for a second generative step, termed the language model, and encodes these as one or more feature vectors. This results in a decoded sequence of words or sub-words based on a specified vocabulary. Recent breakthroughs in object identification and machine translation have influenced significant innovations in image captioning.

The task of image captioning involves two main aspects:
(a) addressing the object detection problem in computer vision and
(b) developing a language model capable of properly generating a statement that describes the identified objects.

One of the hot topics in contemporary machine learning and artificial intelligence is the image caption. It has a large assortment of demands. It can be used in self-driving cars, automatic question answering systems, automatic video description systems, describing a medical situation to someone who is not a doctor, and so on. Furthermore, more than 100 million people worldwide are visually impaired. As a result, describing the current scene can assist them in moving safely.

The following is our contribution to this chapter:

- We compared the LSTM and BERT (large) models and fine-tuned these models to enhance their performances.

- Then we developed a new model for the image captioning system that comprises the concatenation of the BERT model with LSTM and the dense model.

The following is a breakdown of the chapter's structure. In section 12.2, the trending research on image captioning systems is presented. In section 12.3, the proposed workflow and model architecture is described. In section 12.4, our model configuration is discussed. Section 12.5 provides our experiments and result analyses. The constraints and remarks of our model is described in section 12.6.

12.2 LITERATURE REVIEW

The early neural models for image captioning systems were based on template methods. In this method, the essential features are extracted first by using different types of classifiers such as SVM (Support Vector Machine), Naive Bayes, Decision Tree, K-Nearest Neighbour, etc. According to Farhadi et al., the object, activities of the object and the whereabouts of the object are the main features of an object [1]. Then the acquired attributes are converted into textual form. These methods relied on rigid and fixed rules. For a particular situation, an individual set of rules is defined. That pattern can simply reflect the paucity of rules, which affects sentence versatility.

Encoder-Decoder based image captioning models solve this problem efficiently. They have become one of the most widely used frameworks for image captioning systems. The input image is fed into a CNN model in the encoder, which extracts the image's internal representations using feature vectors. After that, a decoder receives the feature vector. At each time step, the decoder creates descriptive captions using the information retrieved from the CNN at a one-word rate.

The attention-based approach was later proposed for the image captioning system. Vaswani et al. recommended a stacked attention architecture [2]. Anderson et al. suggested a combination of top-down and bottom-up visual attention mechanisms. They have used the Faster R-CNN model to encode the input [3]. Lun Huang applied attention to both the encoder and decoder model. In the encoder, the attention mechanism is used to observe the alliances between the objects of the image and in the decoder, attention is used to filter out the unnecessary captions [4].

Transformer architectures have played an outstanding role in a variety of NLP tasks, including computational linguistics, generation of texts, speech recognition, etc. X.Liu used deeper Transformers, which consist of 60 encoder layers and 12 decoder layers, to translate from English to German and English to French [5]. Gu et al. presented a transformer-based non-autoregressive translation model. In their models, the dependency on the previous tokens (target tokens) was eliminated [6]. Zhang et al. integrated the Transformer model alongside some rules and regulations for paraphrasing to make the sentence simple and easy [7]. Devlin et al. designed the BERT language representation framework, which accomplished new SOTA (state of the art) results on 11 NLP tasks [8].

12.3   MODEL ARCHITECTURE

We have employed an encoder-decoder architecture to create our image caption generator system. It is also called a CNN-RNN model.

- The encoder utilizes properties from the image. CNN models are used as encoder.

- The decoder obtains information from CNN in order to provide an image description. RNN models are used as decoder.

In Figure 12.1, an overview of base image caption generator is illustrated.

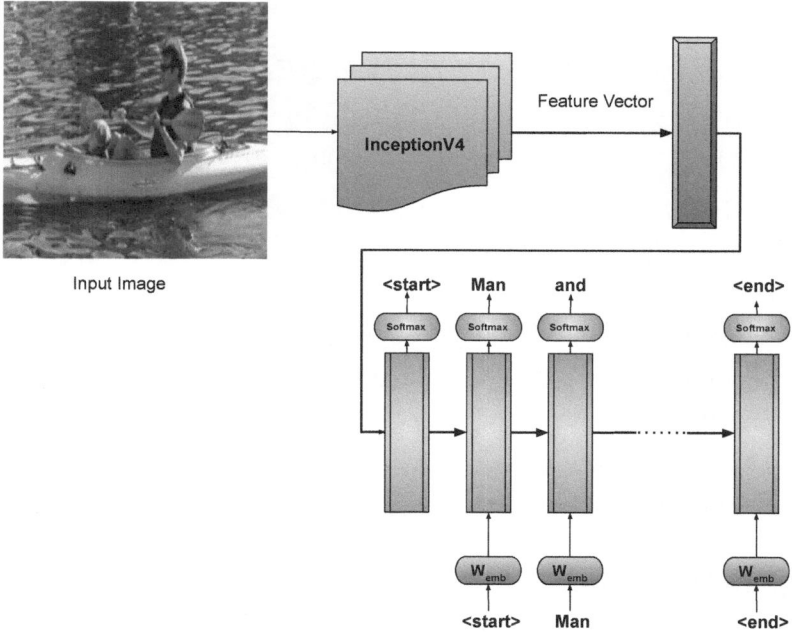

Figure 12.1   The base model image caption architecture

12.3.1   Encoder

We have used InceptionV4 as the encoder. This architecture contains 22 deep layers, and including pooling layers, it is made up of 27 layers. It can classify 1000 categories of images. The input is an image which consists of 299 by 299 pixels. We removed the last layer (fully connected layer) as we only needed to extract the features and used the SoftMax layer as our last layer. We did not perform any fine-tuning.

12.3.2   Decoder

Our experimentation is done with three models, leveraging the LSTM model, the BERT model, and the BERT with concatenated LSTM models.

12.3.2.1 Model-1: Base Model (LSTM: Long-Short Term Memory)

We utilized an LSTM, which is an upgraded version of an RNN (Recurrent Neural Network), to produce caption words one at a time, based on hidden state, context vector, and previously generated words.

In the following two decoders, we used BERT extensions to our base model to enhance its performance.

12.3.2.2 Model-2: Transformer Model (BERT Integration)

BERT (Bidirectional Encoder Representations from Transformers) is a deep bidirectional representation based on both left and right context conditioning.

BERT has two distinct models, $BERT_{base}$ and $BERT_{large}$.

- $BERT_{base}$:Number of transformer blocks (L): 12, hidden units (H): 768, attention heads(A): 12 and parameters: 110 million
- $BERT_{large}$:Number of transformer blocks (L): 24, hidden units (H): 1024, attention heads(A): 16 and parameters: 340 million

We leveraged $BERT large$ to construct the caption's context-dependent word vectors throughout our implementation. To do so, we have done the following tasks:

1. The texts are converted to lower-case and then tokenized using WordPiece and a vocabulary size of 30,000.

2. We have used the [SEP] token to indicate the end of a sentence and the [CLS] token at the beginning of our text.

12.3.2.3 Model-3: Our Model (BERT with LSTM and dense layer)

We amalgamated the Transformer model (BERT) with the LSTM model. Our model is made up of a BERT layer. After that, there's a dropout layer. The output of this layer is fed into two distinct LSTM models, and then the outputs of these two LSTM models are concatenated. And finally, this concatenated layer is followed by several dropout and dense layers.

The structure is shown in Figure 12.2.

12.4   EXPERIMENTAL SETUP

To complete the task, we used the Kaggle environment, which provides a cloud-based service for data science. The setup is as follows:

- 13 GB RAM
- 4.9 GB Hard-Disc
- 16 GB of graphics memory

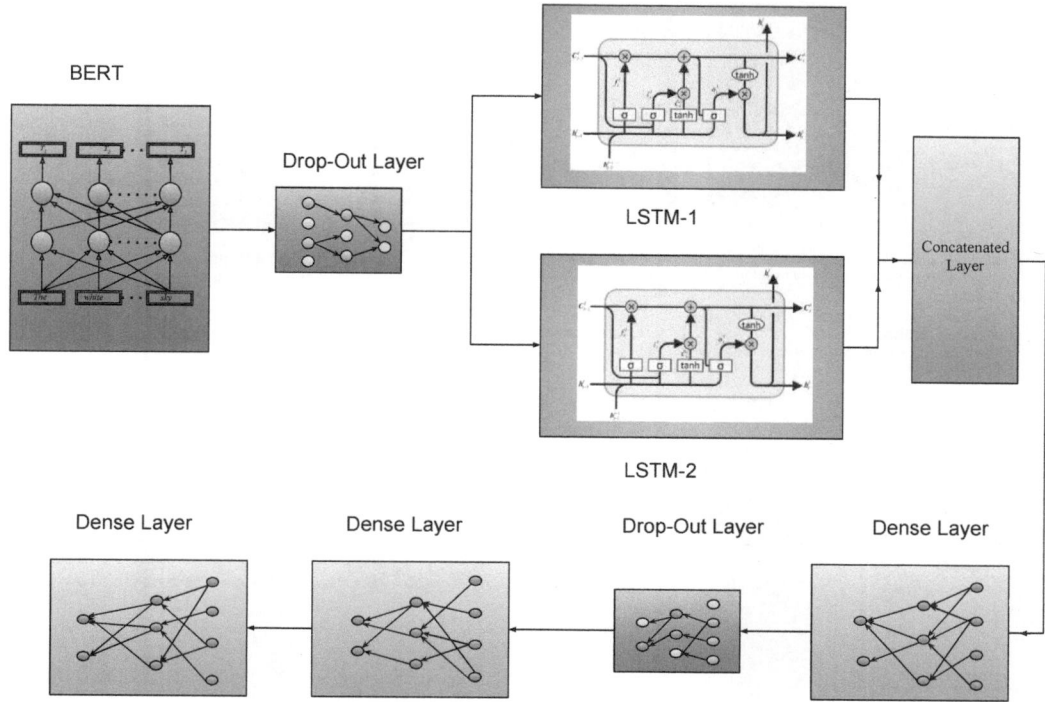

Figure 12.2 Structure of our model

12.4.1 Dataset

Microsoft Common Object in Context (MS-COCO) is one of the largest image datasets[9]. This dataset is used for image recognition, image classification problems, image segmentation problems, and also image captioning problems. Here, almost 2 million images are recorded under 80 different object categories. For every image, there are five individual captions and more than 250,000 people having key points.

12.4.2 Hyper-parameters

During the training phase, we used the following hyper-parameters:

- gradient clip = 3

- batch size = 64

- learning rate of decoder = 0.0005

- dropout rate = 0.4

- epoch= 300

Candidate caption:
The kids play in the wooded area near the water.

Generated caption:
Three kids are playing near the water.

Candidate caption:
Group of young kids play in the water on sunny day.

Generated caption:
Children and adults play with the water.

Candidate caption:
Several young people sitting on rail above crowded beach.

Generated caption:
Group of people sit on wall at the beach.

(a) Examples of accurate captions generated by base model.

Candidate caption:
People stare at the orange fish.

Generated caption:
Some children watching fish in pool.

Candidate caption:
Little leaguer getting ready for pitch.

Generated caption:
Boy in white plays baseball.

Candidate caption:
Young woman in red sequined costume and feather stands on the sidewalk.

Generated caption:
Woman wearing red costume looks at two other people standing on street.

(b) Examples of inaccurate captions generated by base model.

Figure 12.3 Examples of Image Captioning outputs generated by Encoder-Decoder Base model.

12.5 RESULT ANALYSIS

Both qualitative and quantitative analyses have been performed to interpret the results in the following segments.

12.5.1 Qualitative Analysis

Qualitative analysis gives details of the presence or non-appearance of different components in an unknown sample. It uses subjective judgment to analyse the model's performance based on non-quantifiable information.

12.5.1.1 Model-1: Base Model (LSTM: Long-Short Term Memory)

In Figure 12.3, the base model's results have been analysed.

Candidate caption:
Two dogs playing in the sand at the beach.

Generated caption:
Two dogs playing together on a beach.

Candidate caption:
Several people are taking a break while on a snowmobile ride.

Generated caption:
Three people are taking a break while on a snowmobile ride.

Candidate caption:
People watching hot air balloons.

Generated caption:
Crowd watching air balloons at night.

(a) Examples of accurate captions generated by BERT model.

Candidate caption:
Sev ral h k rs rest with th ir g ar in front of mont ain.

Generated caption:
Grop f p op e lay on th dry g ound.

Candidate caption:
Korean man sells soda.

Generated caption:
A man is sitting n front of soda cans.

Candidate caption:
Th you ng an with h wh te tsh rt and jeans is rap ling h rockv all.

Generated caption:
Person with a h lmet and p þ e p nts is rock limb ng

(b) Examples of inaccurate captions generated by BERT model.

Figure 12.4 Examples of Image Captioning outputs generated by BERT model.

We discovered that this model could not produce sentences with the same meaning as the reference sentences while utilizing different terms; it seems that the model is trying to mimic the reference sentence word for word. This may be explained by the fact that this model did not employ any pre-trained embeddings. As a result, the model struggled to acquire reliable word representations that would enable it to transition between comparable terms.

12.5.1.2 Model-2: Transformer Model (BERT Integration)

In Figure 12.4, the BERT integration's results have been analysed. In Figure 12.4a(left), the model has successfully recognized both of the dogs. The face of one dog is not clear in the image, but the model has not faced any difficulties to classify it perfectly. In Figure 12.4a(mid), the model has successfully understood the journey interval. It has also recognized the context of the image. In Figure 12.4a(right), though the elements of the picture are not clearly visible, still the model has distinctly identified the crowds and their activities.

If we go through the inaccurate captions in Figure 12.4b, we can easily understand that though the candidate captions are not exactly the same as the reference captions,

Candidate caption:
Skier is overlooking snow-covered mountain.

Generated caption:
Base: Skier is overlooking snow-covered landscape.
BERT: Skier is overlooking the beautiful white snow covered landscape.
Ours: Hiker standing high on bluff overlooking the mountains below.

Candidate caption:
Black and white dog jumps after yellow toy.

Generated caption:
Base: Dog jumps to catch toy.
BERT: Black dog is jumping up to catch purple and green toy.
Ours: Black and white dog jumps to get the frisbee.

Candidate caption:
Boy jumping in fountain.

Generated caption:
Base: Little boy playing in the water.
BERT: Boy plays in the fountains.
Ours: A young boy is jumping in fountain.

(a) Examples of accurate captions generated by our model.

  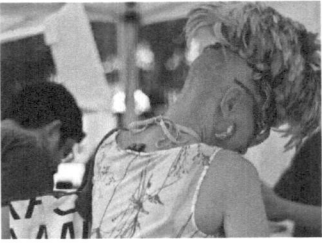

Candidate caption:
Young child is swung by his or her hands while another child sits on grass watching.

Generated caption:
Base: The little girl is being swung around by her arms.
BERT: Little girl in sweater is swung around by an unseen hand.
Ours: Child is sprawled underneath blanket in midair.

Candidate caption:
Girl sitting in dark bar.

Generated caption:
Base: Dark room with chairs.
BERT: Girl sits on bar stool.
Ours: There several people in dark bartype room.

Candidate caption:
Woman with crazy hairdo is shopping outside

Generated caption:
Base: Man in feather hat looking down.
BERT: The person wearing earrings is wearing feathered hat.
Ours: Woman in floral print dress and shaved head at store.

(b) Examples of inaccurate captions generated by our model.

Figure 12.5 Examples of Image Captioning outputs generated by our model.

they are grammatically and contextually correct, which is the objective of this task. BERT captions were typically well-written and had minimal redundancy.

12.5.1.3 *Model-3: Our Model (BERT with LSTM and dense layer)*

Figure 12.5 shows a clear contrast between the three models we've used. The results strongly indicate that our model outperforms them by large margins, though sometimes the BERT model and our models yield similar results.

Our model was able to construct captions with a different writing style from the reference captions by using diverse phrases with similar meanings. This change might be justified by the fact that embeddings stimulate the model to choose the best expression from a set of related words. Figure 12.5a depicts an illustration where

Table 12.1 Comparison of different models

Model	Evaluation Matrix			
	BL1	BL4	METEOR	CIDEr
Attributes & External Knowledge [13]	73	31	25	92
DR with Embedding Reward [14]	71.3	30.4	25.1	93.7
UPDOWN via personality [15]	79.3	36.4	-	124.0
GCN-LSTM+HIP [16]	-	38.0	28.6	120.3
SGAE [17]	81.0	39.0	28.4	129.1
M^2 Transformer [18]	80.8	39.1	29.2	134.5
Ours	**81.21**	**39.5**	**30.89**	**136.7**

the model has used the word "bluff" instead of a "high cliff". "Toy" was translated to "frisbee" in the generated captions in the middle image in Figure 12.5a.

The captions, which are marked as inaccurate in Figure 12.5b, also describe the image precisely. The generated sentences are error-free. In short, our model is very accurate in representing contextualized words.

12.5.2 Quantitative Analysis

We have used three evaluation metrics (BLEU, METEOR, and CIDEr) to evaluate the performance of our model.

The BLEU score compares a sentence against one or more reference sentences and tells how well the candidate sentence matches the list of reference sentences. It gives an output score of between 0 and 1 [10].

The METEOR automatic assessment metric evaluates them by matching machine translation hypotheses to one or more reference translations, the METEOR automatic assessment metric evaluates them. METEOR was created to specifically solve the above-mentioned flaws in BLEU. It calculates a score based on explicit word-to-word matches between the translation and a reference translation when evaluating a translation. If more than one reference translation is available, the supplied translation is assessed against each reference independently, and the best score is recorded. [11].

CIDEr (Consensus-based Image Description Evaluation) compares the resemblance of a produced sentence to a set of human-written ground truth sentences [12].

In Table 12.1, we have compared the BLEU (BL1 & BL4), METEOR and CIDEr scores of our BERT integrated model with the existing models.

12.6 CONCLUSION

From our experiments, we conclude that for a very small corpora, the base LSTM model shows a better output, but for the large corpora, our model surpluses all the existing models, whereas, the BERT model shows a constant output for both the small and large corpora.

The main limitation of our model is that all captions must be less than 50 words. It doesn't provide a way to properly represent an image across multiple lines or paragraphs. Second, our framework cannot account for concept art or cartoons. It cannot generate metaphorical captions. Subtitles aren't always great. Sometimes they are just reasonable. Also, it cannot distinguish between specific places or celebrities.

In the future, we will implement our model in connected images (images of the same events or albums). Moreover, the effect on abstract or metaphoric images will also be inspected.

Bibliography

[1] Farhadi, Ali, Mohsen Hejrati, Mohammad Amin Sadeghi, Peter Young, Cyrus Rashtchian, Julia Hockenmaier, and David Forsyth. "Every picture tells a story: Generating sentences from images." In *European conference on computer vision*, pp. 15-29. Springer, Berlin, Heidelberg, 2010.

[2] Vaswani, Ashish, Noam Shazeer, Niki Parmar, Jakob Uszkoreit, Llion Jones, Aidan N. Gomez, Lukasz Kaiser, and Illia Polosukhin. "Attention is all you need." *Advances in neural information processing systems* 30 (2017).

[3] Anderson, Peter, Xiaodong He, Chris Buehler, Damien Teney, Mark Johnson, Stephen Gould, and Lei Zhang. "Bottom-up and top-down attention for image captioning and visual question answering." *In Proceedings of the IEEE conference on computer vision and pattern recognition*, pp. 6077-6086. 2018.

[4] Huang, Lun, Wenmin Wang, Jie Chen, and Xiao-Yong Wei. "Attention on attention for image captioning." *In Proceedings of the IEEE/CVF international conference on computer vision*, pp. 4634-4643. 2019.

[5] Liu, Xiaodong, Kevin Duh, Liyuan Liu, and Jianfeng Gao. "Very deep transformers for neural machine translation." *arXiv preprint arXiv:2008.07772)* (2020).

[6] Gu, Jiuxiang, Zhenhua Wang, Jason Kuen, Lianyang Ma, Amir Shahroudy, Bing Shuai, Ting Liu et al. "Recent advances in convolutional neural networks." *Pattern recognition* 77 (2018): 354-377.

[7] Zhang, Jiacheng, Huanbo Luan, Maosong Sun, Feifei Zhai, Jingfang Xu, Min Zhang, and Yang Liu. "Improving the transformer translation model with document-level context." *arXiv preprint arXiv:1810.03581* (2018).

[8] Devlin, Jacob, Ming-Wei Chang, Kenton Lee, and Kristina Toutanova. "Bert: Pre-training of deep bidirectional transformers for language understanding." *arXiv preprint arXiv:1810.04805* (2018).

[9] Lin, Tsung-Yi, Michael Maire, Serge Belongie, James Hays, Pietro Perona, Deva Ramanan, Piotr Dollár, and C. Lawrence Zitnick. "Microsoft coco: Common

objects in context." In *European conference on computer vision*, pp. 740-755. Springer, Cham, 2014.

[10] Papineni, Kishore, Salim Roukos, Todd Ward, and Wei-Jing Zhu. "Bleu: a method for automatic evaluation of machine translation." In *Proceedings of the 40th annual meeting of the association for computational linguistics*, pp. 311-318. 2002.

[11] Banerjee, Satanjeev, and Alon Lavie. "METEOR: An automatic metric for MT evaluation with improved correlation with human judgments." In *Proceedings of the ACL workshop on intrinsic and extrinsic evaluation measures for machine translation and/or summarization*, pp. 65-72. 2005.

[12] Vedantam, Ramakrishna, C. Lawrence Zitnick, and Devi Parikh. "Cider: Consensus-based image description evaluation." In *Proceedings of the IEEE conference on computer vision and pattern recognition*, pp. 4566-4575. 2015.

[13] Wu, Qi, Chunhua Shen, Peng Wang, Anthony Dick, and Anton Van Den Hengel. "Image captioning and visual question answering based on attributes and external knowledge." *IEEE transactions on pattern analysis and machine intelligence* 40, no. 6 (2017): 1367-1381.

[14] Ren, Zhou, Xiaoyu Wang, Ning Zhang, Xutao Lv, and Li-Jia Li. "Deep reinforcement learning-based image captioning with embedding reward." In *Proceedings of the IEEE conference on computer vision and pattern recognition*, pp. 290-298. 2017.

[15] Shuster, Kurt, Samuel Humeau, Hexiang Hu, Antoine Bordes, and Jason Weston. "Engaging image captioning via personality." In *Proceedings of the IEEE/CVF conference on computer vision and pattern recognition*, pp. 12516-12526. 2019.

[16] Yao, Ting, Yingwei Pan, Yehao Li, and Tao Mei. "Hierarchy parsing for image captioning." In *Proceedings of the IEEE/CVF international conference on computer vision*, pp. 2621-2629. 2019.

[17] Yang, Xu, Kaihua Tang, Hanwang Zhang, and Jianfei Cai. "Auto-encoding scene graphs for image captioning." In *Proceedings of the IEEE/CVF conference on computer vision and pattern recognition*, pp. 10685-10694. 2019.

[18] Cornia, Marcella, Matteo Stefanini, Lorenzo Baraldi, and Rita Cucchiara. "Meshed-memory transformer for image captioning." In *Proceedings of the IEEE/CVF conference on computer vision and pattern recognition*, pp. 10578-10587. 2020.

CHAPTER 13

A Convolutional Neural Network-based Approach to Recognize Bangla Handwritten Characters

Mohammad Golam Mortuza

Dept. of Electronics and Telecommunication Engineering, Chittagong University of Engineering and Technology, Chittagong, Bangladesh

Saiful Islam

Dept. of Electronics and Telecommunication Engineering, Chittagong University of Engineering and Technology, Chittagong, Bangladesh

Md. Humayun Kabir

Dept. of Electronics and Telecommunication Engineering, Chittagong University of Engineering and Technology, Chittagong, Bangladesh

Uipil Chong

University of Ulsan, South Korea

CONTENTS

Bangla is one of the most frequently used and spoken languages of the world but not much research has been conducted on recognizing Bangla handwritten characters. Almost all government and non-government offices are being digitized nowadays and their previous records need to be digitized too. It is a very tough and dismal human job to convert all the records manually. So, there is a keen need to develop a system that will automatically detect and recognize all the Bangla characters in a document and generate a digital copy of them. This research proposes a Convolution Neural Network based approach to recognize handwritten Bangla characters. The primary goal of this research is not just to create a recognition system, but rather to obtain a CNN-based systematic approach to recognizing Bangla handwritten characters. This research focuses on improving the accuracy of character recognition and draws a comparative analysis of validation accuracy between different character classes. For this experiment, the Banglalekha-Isolated dataset was used which is an 84-class dataset and contains 1,66,105 handwritten characters of vowels, consonants, numbers, and special joint letters in Bangla. The ICT division, which is a wing of the Ministry of ICT, Bangladesh, has funded this collection of datasets and made it publicly available for all sorts of research. The validation accuracy achieved on the full dataset is 92.02%, 97.58% was achieved for 11 vowel classes, 94.70% for 39 consonant classes, 98.00% for 10 number classes, and 90.03% for 24 joint letter classes. Also, the validation accuracy and loss were later compared with another contemporary research on Bangla handwritten character recognition system that used the same Banglalekha-Isolated dataset.

13.1   INTRODUCTION

Bangla is the mother language of Bangladesh. After Hindi, Bangla is the second most famous and practiced language of the twenty-two organized languages of India[1]. As a native language, Bangla is the fifth most spoken and by the total number of speakers in the world, Bangla is the seventh most-spoken language with approximately 228 million native speakers and another 37 million as second language speakers [2]. Bangla is the national and official language of Bangladesh. Bangla is used by 98% of Bangladeshis as their first language [1]. West Bengal, the Barak Valley region (which is in the state of Assam), and the Tripura state of India use Bangla as their official language. Also, Bangla is widely practiced by the Bengali community across the world. From 1948 to 1956, the Bengali language movement demanded Bangla be the official language of Pakistan then developed Bengali nationalism in East Pakistan paved the pathway of freedom of Bangladesh in 1971. UNESCO recognized 21 February as International Mother Language Day in 1999 in recognition of the language movement [3]. There are 50 basic alphabets in the Bangla language, amongst which 11 are vowels and 39 are consonants. The Bengali language ties together a culturally diverse region and it is the classic component of Bengali identity.

Despite being a broadly popular language, not so much research has been conducted on recognizing handwritten characters of the Bangla language compared to Chinese, English, and Arabic language [4]. For analyzing visual imagery, the Convolution Neural Network (CNN) is most commonly applied, which is a class of deep neural networks [5][6]. CNN's are based on the principle of multi-layer perceptions, which means fully connected networks [7][20]. All the neurons of the next layer are connected with each neuron of the previous layer in Fully Connected networks [8][21]. The CNN consists of 3 layers – (a) input layer, (b) hidden layer, and (c) output layer. Normally, the hidden layer consists of several layers that perform the dot product, the most commonly used activation function is the Rectified Linear Unit (ReLU) followed by other layers such as pooling layers, fully connected layers, and normalization layers[22].

This research focuses on training and recognizing handwriting characters of the Bangla language. The goals followed for this research are to create a model to train the dataset of Bangla handwritten characters, i.e., the Banglalekha-Isolated dataset [9] in different sizes individually, that is, only 10 digits, 11 vowels, 39 consonants, 24 joint alphabets and finally the entire dataset of 84 handwritten Bangla characters. Next, drawing a comparison analysis between the validation accuracy achieved in individual cases. The rest of this article is organized as follows: In section 13.2, we mention some of the significant contributions done in the field of Bangla handwritten character recognition. In section 13.3, we describe the methodology and system architecture used for this research. Section 13.4 describes the dataset used for this research. Section 13.5 covers the analysis of the results of this study. And finally, section 13.6 ends the chapter and talks about some future extents of progress.

13.2 RELATED WORK

A massive amount of research work has been done on recognizing the handwritten characters of Chinese, English, and Arabic languages[4], but not much research has been conducted on the Bengali language. But in the Indian subcontinent, there have been some notable research works done by data scientists and machine learning experts.

Research concentrating on the Bangla character recognition system for printed characters has been done by Md. Mahbub Alam et al. in 2010[10]. They used the MATLAB software v7.0 for their experiment. Sutonny, sulekha, sunetra, etc, fonts were used in their experiment. They observed that if the font size was enlarged, they could obtain better results. They obtained an accuracy of about 97%.

Md. Mahbub Rahman et al. investigated research on CNN (convolution neural network) based Bangla handwritten character recognition[11]. The written character images are normalized first in their proposed method and then the Convolution Neural Network is used for classifying individual characters. For this study, 20000 handwritten characters of different variants were used. Out of 2500 test cases, 351 cases were misclassified by their proposed BHCR-CNN, and thus it achieved an accuracy of 85.96%.

In 2017, Biswajit Purkaystha et al. proposed a convolutional deep model for recognizing handwritten Bangla characters[12]. In their model, by using kernels and locally respected fields, they first extracted the features, and then, for the determining task, they used densely connected layers. Their method has been tested on the Banglalekha-Isolated dataset. They achieved an accuracy of 91.60% on compound letters, 94.99% on vowels, 91.23% on alphabets, 98.66% on numerals, and 89.93 percent on almost all Bangla characters.

Mithun Biswas et al. presented a dataset of Bangla handwritten isolated characters in an article in 2017[9]. A total of 84 different characters were contained in that dataset. Among these are 10 Bangla numerals, 50 Bangla basic characters, and 24 selected compound characters. A collection of 2000 handwriting samples for each of the

84 characters have been collected by them, then digitized, and finally pre-processed for further electronic usage. In the final dataset, 1,66,105 handwritten character images were included after discarding the scribbles and mistakes.

In 2018, AKM Shahriar Azad Rabby et al. published an article where they propose a simple and lightweight CNN model for classifying handwritten Bangla characters (39 consonants and 11 vowels)[13]. They made experiments on three datasets i.e., Banglalekha-Isolated, CMATERdb, and the ISI dataset. The BornoNet model that they proposed, achieved 96.40%, 95.71%, 96.81%, and 98% validation accuracy, respectively, for the mixed, Banglalekha-Isolated, ISI and CMATERdb. So far for ISI, CMATERdb, and Banglalekha-Isolated datasets, their proposed model has achieved the best accuracy rate[13].

An article was published by Md. Zahangir Alom et al. in 2018 discussed deep convolutional neural networks and their performance was systematically evaluated on the application of recognizing handwritten Bangla characters[14]. According to their claim, based on the output results of their experiment, DenseNet has the best performance in classifying Bangla alphabets, special characters, and digits. Using DenseNet, they achieved a recognition rate of 98.31%, 99.13%, and 98.18% for handwritten Bangla alphabets, digits, and special characters respectively.

In 2019, Rumman Rashid Chowdhury et al. proposed a process of recognizing handwritten characters and converting them to electronically editable format [15]. By using the Convolution Neural Network on the base dataset of the Banglalekha-Isolated, they achieved 91.81% accuracy on alphabets. Then, by using augmentation of data, they expanded the number of images to 200,000 and achieved an accuracy of 95.25%. The model was also hosted on a web server for further access and usage.

13.3 METHODOLOGY AND SYSTEM ARCHITECTURE

CNN was influenced by the principle of the visual cortex of the human brain[7]. Complex features are detected by the human brain through the layers of the cortex and then the brain recognizes what was seen. In CNN architecture, the same thing is done by applying some layers and processing categories[8] [12]. The model developed for recognizing handwritten Bangla characters was implemented using the MATLAB[16] software, v2020a. Gray scale images having the size of]32x32] have been used as input in the system with 'zero center' normalization. The first convolution layer contains 64 filters and a kernel size of [3x3], batch normalization with 64 channels having activations of [32x32x64] is done, and then Rectified Linear Unit (ReLU)[17] is used as the activation function in this layer. Then a Maxpooling layer is used afterward that has a pool size of [2x2].

Then another convolution layer is used having 128 filters with a kernel size of [3x3]. Then there is batch normalization done with 128 channels and the ReLU activation function used which has the size of [16x16x128]. This layer is also followed by a Maxpooling layer having stride [2x2]. After that, another convolution layer is used having 256 filters with a kernel size of [3x3]. Then there is batch normalization done with 256 channels and the ReLU activation function used which has the size of [8x8x256]. A Maxpooling layer follows this layer having stride [2x2], which converts

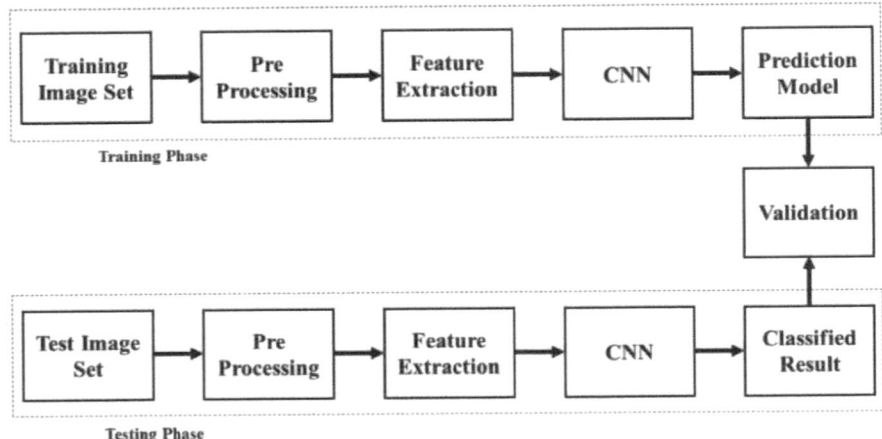

Figure 13.1 Block diagram of the proposed method

the size of the activations to [4x4x256]. Then another final convolution layer is introduced in the system having 512 filters with a kernel size of [3x3]. Then there is batch normalization done with 512 channels and the ReLU activation function used which has the size of [4x4x512]. A Maxpooling layer follows this layer having stride [2x2] that converts the size of the activations to [2x2x512].

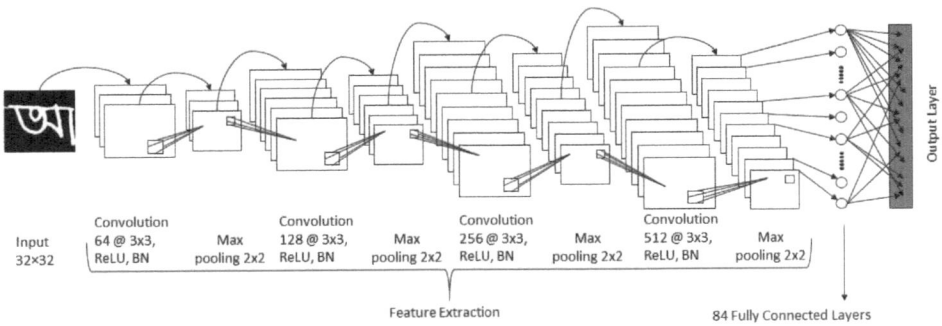

Figure 13.2 Proposed model of CNN architecture

Then, Fully Connected Layers are introduced that connect 84 layers for the full dataset, 10 layers for the numbers, 11 layers for the vowels, 39 layers for the consonants, and 24 layers for the joint alphabets, respectively. For output layers for each of them, a Softmax activation function[18] is used for providing probabilistic values for every class. Then the classifier model is assembled, and Stochastic Gradient Descent (SGDM) is used as the output optimization function for this model[19] with a learning rate of 0.01.

13.4   DATASET

The Banglalekha-Isolated dataset has been used in this experiment [9]. It is a collection of Bangla handwritten isolated character samples. It contains 24 compound characters, 10 Bangla numerals, and 50 Bangla basic characters. A total of 2000 samples of handwriting for each of the 84 characters were collected, preprocessed, and digitized. A total of 1,66,105 final handwritten character images were added to the final dataset after discarding the scribbles and mistakes. The ICT division, which is a wing of the Ministry of ICT, Bangladesh, has funded this collection of datasets and made it publicly available for all sorts of research [9].

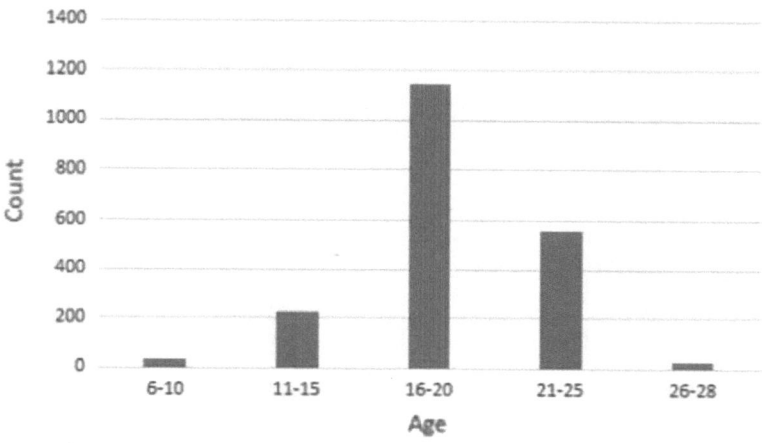

Figure 13.3   Age distribution of the subjects [9].

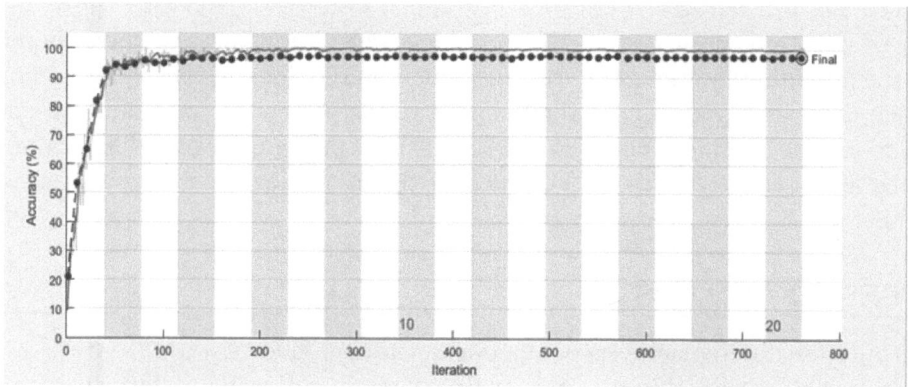

Figure 13.4   Validation accuracy of the vowel class.

Among the 1,66,105 images from the original dataset, randomly 600 images from each of the 84 classes were chosen and converted to the specific size of [32x32x1] as part of prepossessing. The new database contains 50,400 images of 84 classes 600 images in each class. Randomly, 450 images (75%) from each class were selected for training the model, and 150 images (25%) from each class for validation. The number

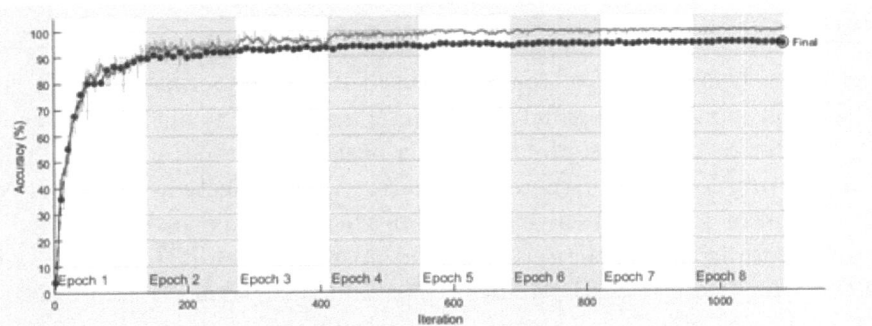

Figure 13.5 Validation accuracy of the consonant class.

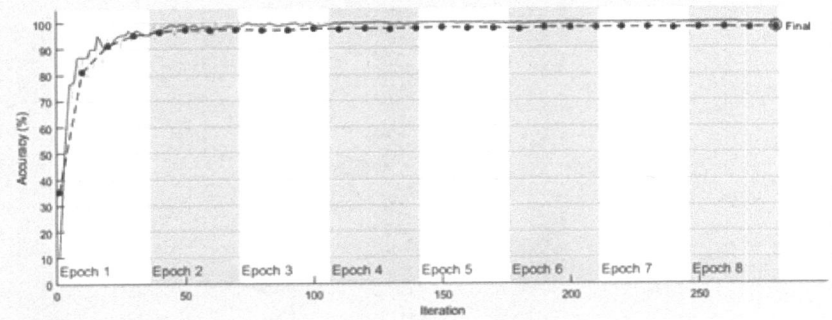

Figure 13.6 Validation accuracy of the number class.

of epochs used is 8, which means, the whole dataset will pass through the model 8 times. For the full dataset, iterations per epoch done were 295 and the maximum number of iterations done were 2360. For the vowels, iterations per epoch done were 38 and the maximum number of iterations done was 760, for the consonants, iterations per epoch done were 295 among 1096, 35 iterations per epoch were done among 280 iterations for the number classes and 84 iterations per epoch were done among 672 iterations for joint letter classes. For each of them, the learning rate was scheduled to be constant.

13.5 RESULT ANALYSIS

The outputs for different classes of the dataset are as follows.

After finishing the training, it is observed that validation accuracy of 97.58% was achieved for the vowel class, 94.70% for the consonant class, 98.00% for the number class, 90.03% for the joint letter class and 92.02% for the full dataset. The error percentage observed was 2.42 percent for vowels, 5.30 percent for consonants, 2.00 percent for numbers, 9.97 percent for joint letters, and 7.98 percent for the full dataset. The validation loss curves for different classes go down very rapidly in early epochs and become more stable afterward and stay very close to zero.

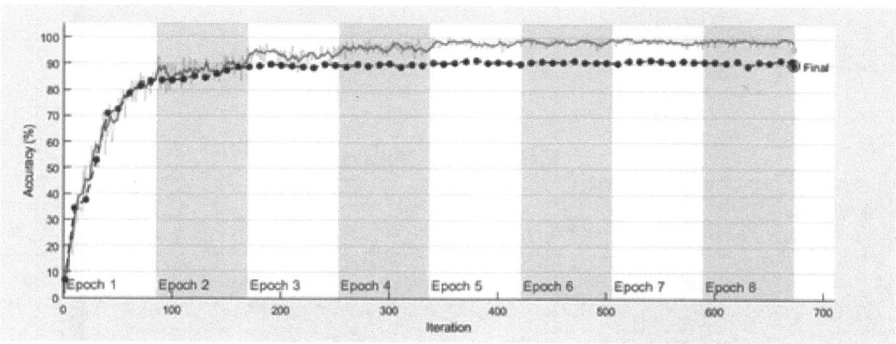

Figure 13.7 Validation accuracy of the joint letter class.

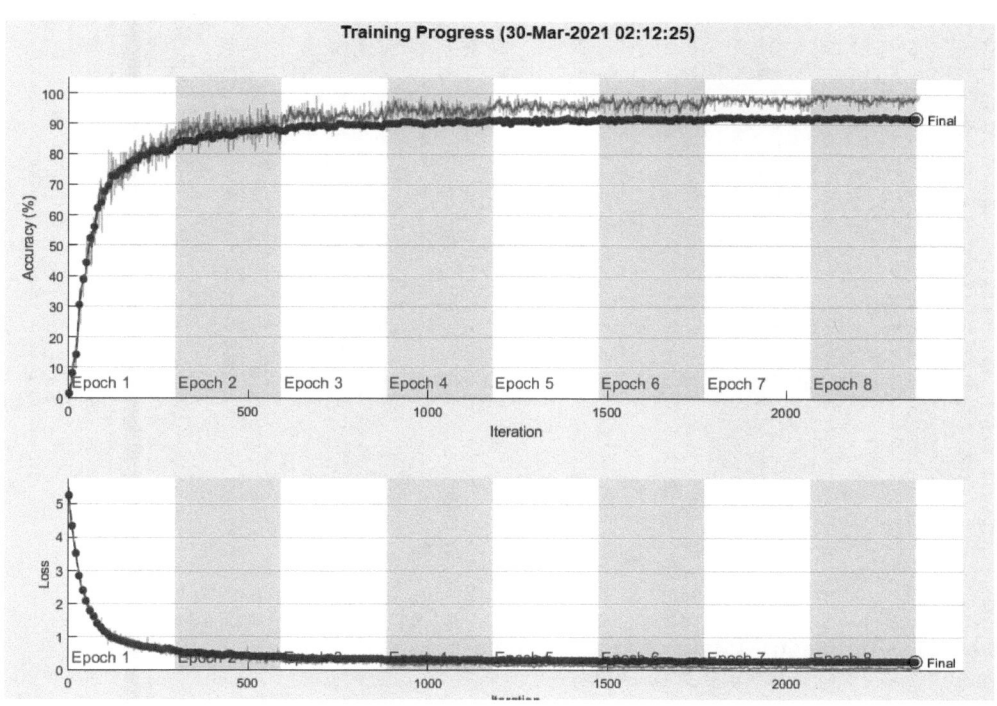

Figure 13.8 Validation accuracy of the full dataset class.

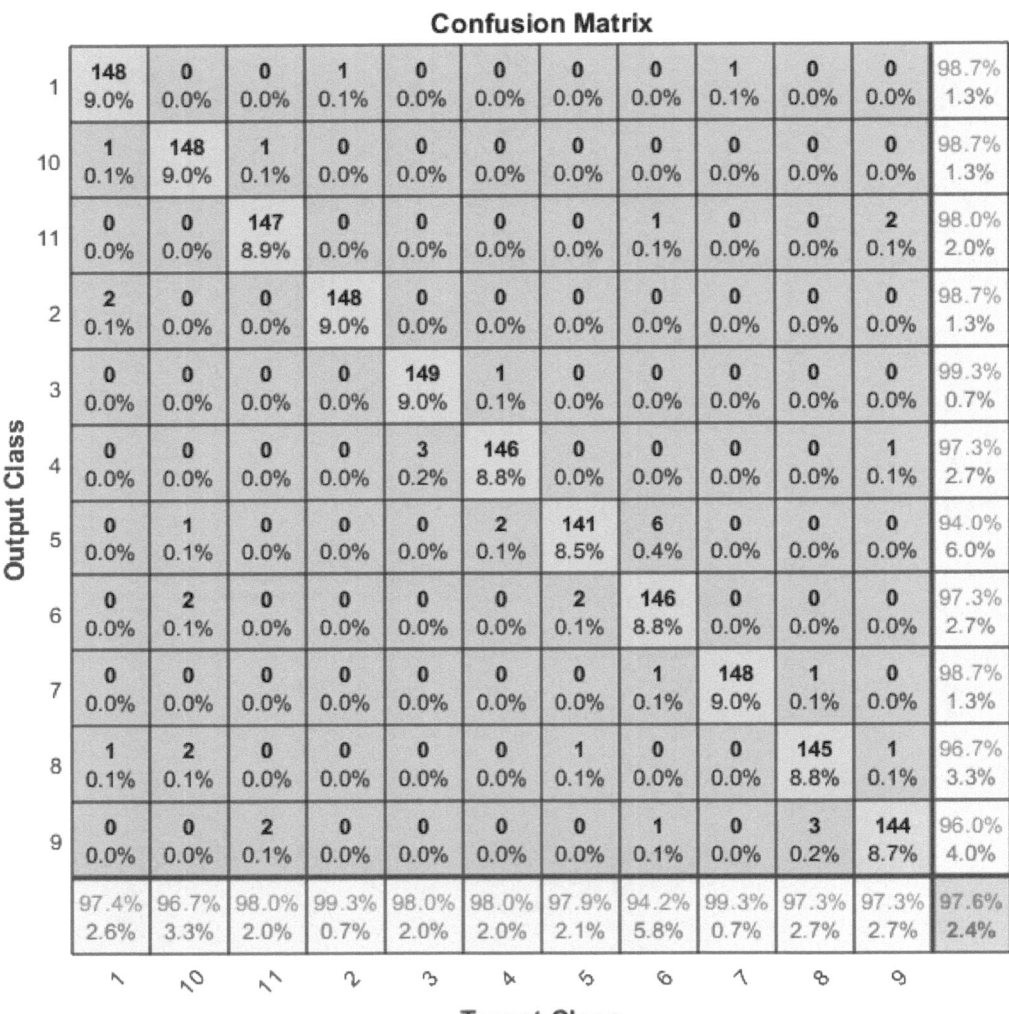

Figure 13.9 Confusion Matrix of the vowel class.

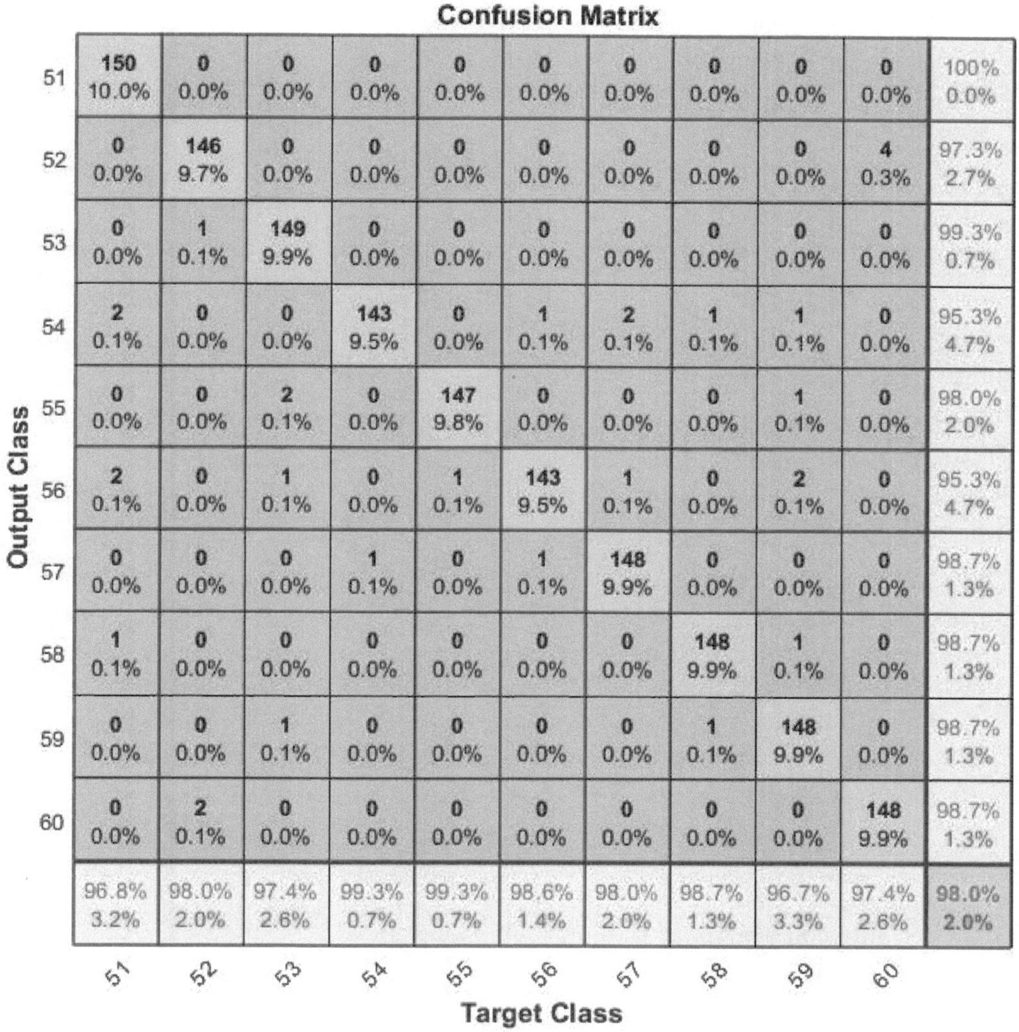

Confusion Matrix

Output Class	51	52	53	54	55	56	57	58	59	60	
51	**150** 10.0%	**0** 0.0%	**0** 0.0%	**0** 0.0%	**0** 0.0%	**0** 0.0%	**0** 0.0%	**0** 0.0%	**0** 0.0%	**0** 0.0%	100% 0.0%
52	**0** 0.0%	**146** 9.7%	**0** 0.0%	**0** 0.0%	**0** 0.0%	**0** 0.0%	**0** 0.0%	**0** 0.0%	**0** 0.0%	**4** 0.3%	97.3% 2.7%
53	**0** 0.0%	**1** 0.1%	**149** 9.9%	**0** 0.0%	**0** 0.0%	**0** 0.0%	**0** 0.0%	**0** 0.0%	**0** 0.0%	**0** 0.0%	99.3% 0.7%
54	**2** 0.1%	**0** 0.0%	**0** 0.0%	**143** 9.5%	**0** 0.0%	**1** 0.1%	**2** 0.1%	**1** 0.1%	**1** 0.1%	**0** 0.0%	95.3% 4.7%
55	**0** 0.0%	**0** 0.0%	**2** 0.1%	**0** 0.0%	**147** 9.8%	**0** 0.0%	**0** 0.0%	**0** 0.0%	**1** 0.1%	**0** 0.0%	98.0% 2.0%
56	**2** 0.1%	**0** 0.0%	**1** 0.1%	**0** 0.0%	**1** 0.1%	**143** 9.5%	**1** 0.1%	**0** 0.0%	**2** 0.1%	**0** 0.0%	95.3% 4.7%
57	**0** 0.0%	**0** 0.0%	**0** 0.0%	**1** 0.1%	**0** 0.0%	**1** 0.1%	**148** 9.9%	**0** 0.0%	**0** 0.0%	**0** 0.0%	98.7% 1.3%
58	**1** 0.1%	**0** 0.0%	**0** 0.0%	**0** 0.0%	**0** 0.0%	**0** 0.0%	**0** 0.0%	**148** 9.9%	**1** 0.1%	**0** 0.0%	98.7% 1.3%
59	**0** 0.0%	**0** 0.0%	**1** 0.1%	**0** 0.0%	**0** 0.0%	**0** 0.0%	**0** 0.0%	**1** 0.1%	**148** 9.9%	**0** 0.0%	98.7% 1.3%
60	**0** 0.0%	**2** 0.1%	**0** 0.0%	**0** 0.0%	**0** 0.0%	**0** 0.0%	**0** 0.0%	**0** 0.0%	**0** 0.0%	**148** 9.9%	98.7% 1.3%
	96.8% 3.2%	98.0% 2.0%	97.4% 2.6%	99.3% 0.7%	99.3% 0.7%	98.6% 1.4%	98.0% 2.0%	98.7% 1.3%	96.7% 3.3%	97.4% 2.6%	98.0% 2.0%

Target Class

Figure 13.10 Confusion Matrix of the number class.

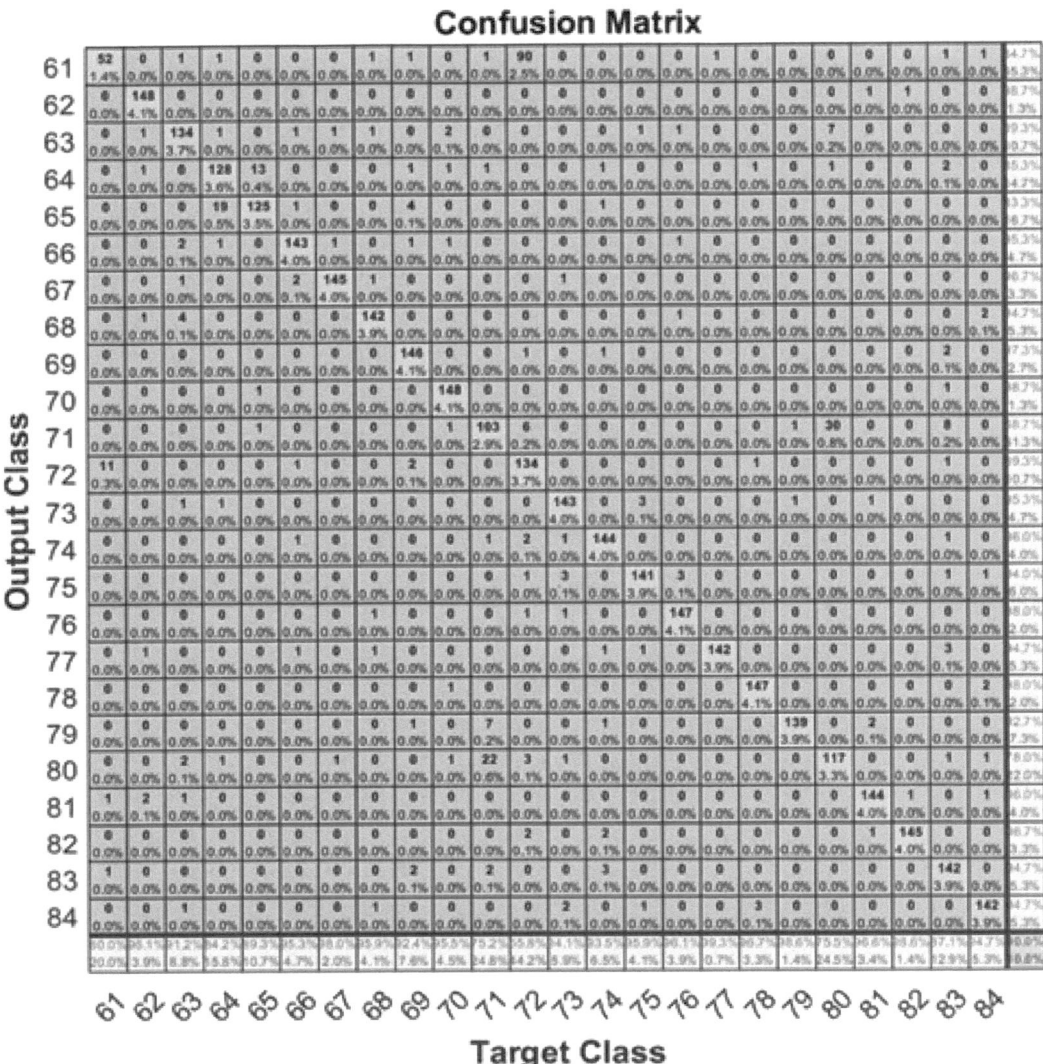

Figure 13.11 Confusion Matrix of the joint letter class.

13.6 CONCLUSION AND FUTURE WORK

This research focused on recognizing handwritten Bangla characters only and observed the effects of CNN in recognizing handwritten characters. Alphabet recognition is a primary step for implementing the model in multiple character recognition and is followed by word, sentence, and paragraph gradually. When recognizing a word that comes forward, first of all, each alphabet should have to be separated and recognized individually which this research completes. So, for the future scope of this research, the authors propose designing a model that will detect long words and gradually recognize long sentences accurately.

Table 13.1 Accuracy of the model for different types of character classes of the BanglaLekha-Isolated[9] dataset.

Type	Classes	Test Accuracy(percentage)	Error(percentage)
Vowels	11	97.58	2.42
Digits	10	98.00	2.00
Consonants	39	94.70	5.30
Joint Letters	24	90.03	9.97
All characters	84	92.02	7.98

Table 13.2 Comparison of Test Accuracy(percentage) of the Proposed System with another contemporary method of handwritten Bangla character recognition that used Banglalekha-Isolated Dataset.

Type	Classes	Proposed System	Bishwajit Purkaystha et al.[12]
Digits	10	98.00	98.66
Vowels	11	97.58	94.99
Consonants	39	94.70	-
Joint Letters	24	90.03	91.60
All characters	84	92.02	89.93

Bibliography

[1] (Bengali language – Wikipedia, 2021)

[2] Ghosh, I. (2020, February 15). Ranked: The 100 Most Spoken Languages Around the World. Visual Capitalist. *https://www.visualcapitalist.com/100-most-spoken-languages/*

[3] UNESCO; "Resolution adopted by the 30th Session of UNESCO's General Conference (1999)," *UNESCO*, 1999.

[4] Mirvaziri, Hamid, Mohammad Masood Javidi, and Najme Mansouri. "Handwriting recognition algorithm in different languages: Survey." In *International Visual Informatics Conference*, pp. 487-497. Springer, Berlin, Heidelberg, 2009.

[5] Li, Gongfa, Heng Tang, Ying Sun, Jianyi Kong, Guozhang Jiang, Du Jiang, Bo Tao, Shuang Xu, and Honghai Liu. "Hand gesture recognition based on convolution neural network." *Cluster Computing 22*, no. 2 (2019): 2719-2729.

[6] Ahmed, Sifat, Fatima Tabsun, Abdus Sayef Reyadh, Asif Imtiaz Shaafi, & Faisal Muhammad Shah. "Bengali handwritten alphabet recognition using deep convolutional neural network." In *2019 International Conference on Computer, Communication, Chemical, Materials and Electronic Engineering (IC4ME2)*, pp. 1-4. IEEE, 2019.

[7] Mo, Weilong, Xiaoshu Luo, Yexiu Zhong, & Wenjie Jiang. "Image recognition using convolutional neural network combined with ensemble learning algorithm."

In *Journal of Physics: Conference Series*, vol. 1237, no. 2, p. 022026. IOP Publishing, 2019.

[8] Tao, Wenjin, Ming C. Leu, & Zhaozheng Yin. "American Sign Language alphabet recognition using convolutional neural networks with multiview augmentation and inference fusion." *Engineering Applications of Artificial Intelligence* 76 (2018): 202-213.

[9] Biswas, Mithun, Rafiqul Islam, Gautam Kumar Shom, Md Shopon, Nabeel Mohammed, Sifat Momen, & Anowarul Abedin. "Banglalekha-isolated: A multipurpose comprehensive dataset of handwritten Bangla isolated characters." *Data in Brief* 12 (2017): 103-107.

[10] Alam, Md Mahbub, & M. Abul Kashem. "A complete Bangla OCR system for printed characters." *JCIT* 1, no. 01 (2010): 30-35.

[11] Rahman, Md Mahbubar, M. A. H. Akhand, Shahidul Islam, Pintu Chandra Shill, & MM Hafizur Rahman. "Bangla handwritten character recognition using convolutional neural network." *International Journal of Image, Graphics and Signal Processing* 7, no. 8 (2015): 42.

[12] Purkaystha, Bishwajit, Tapos Datta, & Md Saiful Islam. "Bengali handwritten character recognition using deep convolutional neural network." In *2017 20th International Conference of Computer and Information Technology (ICCIT)*, pp. 1-5. IEEE, 2017.

[13] Zahangir Alom, Md, Peheding Sidike, Mahmudul Hasan, Tark M. Taha, and Vijayan K. Asari. "Handwritten Bangla character recognition using the state-of-art deep convolutional neural networks." *arXiv e-prints* (2017): arXiv-1712.

[14] Chowdhury, Rumman Rashid, Mohammad Shahadat Hossain, Raihan ul Islam, Karl Andersson, & Sazzad Hossain. "Bangla handwritten character recognition using convolutional neural network with data augmentation." In *2019 Joint 8th International Conference on Informatics, Electronics & Vision (ICIEV) and 2019 3rd International Conference on Imaging, Vision & Pattern Recognition (icIVPR)*, pp. 318-323. IEEE, 2019.

[15] Hazra, Abhishek, Prakash Choudhary, Sanasam Inunganbi, & Mainak Adhikari. "Bangla-Meitei Mayek scripts handwritten character recognition using convolutional neural network." *Applied Intelligence* 51, no. 4 (2021): 2291-2311.

[16] Ghosh, Tapotosh, Hasan Al Banna, Nasirul Mumenin, & Mohammad Abu Yousuf. "Performance analysis of state of the art convolutional neural network architectures in Bangla handwritten character recognition." *Pattern Recognition and Image Analysis* 31, no. 1 (2021): 60-71.

[17] Rabby, AKM Shahariar Azad, Md Majedul Islam, Nazmul Hasan, Jebun Nahar, & Fuad Rahman. "Borno: Bangla handwritten character recognition using a multiclass convolutional neural network." In *Proceedings of the Future Technologies Conference*, pp. 457-472. Springer, Cham, 2020.

[18] Raisa, Jasiya Fairiz, Maliha Ulfat, Abdullah Al Mueed, & Mohammad Abu Yousuf. "Handwritten Bangla character recognition using convolutional neural network and bidirectional long short-term memory." In *Proceedings of International Conference on Trends in Computational and Cognitive Engineering*, pp. 89-101. Springer, Singapore, 2021.

[19] Kibria, Md Raisul, Afrin Ahmed, Zannatul Firdawsi, & Mohammad Abu Yousuf. "Bangla compound character recognition using support vector machine (SVM) on advanced feature sets." In *2020 IEEE Region 10 Symposium (TENSYMP)*, pp. 965-968. IEEE, 2020.

[20] Dhar, Jitu Prakash, Md Saiful Islam, & Muhammad Ahsan Ullah. "A fuzzy logic based contrast and edge sensitive digital image watermarking technique." *SN Applied Sciences* 1, no. 7 (2019): 1-9.

[21] Rahman, Takowa, & Md Saiful Islam. "Image segmentation based on fuzzy C means clustering algorithm and morphological reconstruction." In *2021 International Conference on Information and Communication Technology for Sustainable Development (ICICT4SD)*, pp. 259-263. IEEE, 2021.

[22] Islam, Md Saiful, & Ui Pil Chong. "A digital image watermarking algorithm based on DWT DCT and SVD." *International Journal of Computer and Communication Engineering* 3, no. 5 (2014): 356.

CHAPTER 14

Flood Region Detection Based on K-Means Algorithm and Color Probability

Promiti Chakraborty

Department of Computer Science & Engineering (CSE), Chittagong University of Engineering & Technology(CUET), Chattogram, Bangladesh

Sabiha Anan

Department of Computer Science & Engineering (CSE), Chittagong University of Engineering & Technology(CUET), Chattogram, Bangladesh

Kaushik Deb

Department of Computer Science & Engineering (CSE), Chittagong University of Engineering & Technology(CUET), Chattogram, Bangladesh

CONTENTS

BANGLADESH is an agricultural country. It is susceptible to flood-driven agricultural losses and suffers more during the rainy season. Thus, the afflicted agricultural production directly influences the economy of the country. Therefore, it is essential to determine the severity of the flood for the well-being of agriculture and human life. This is very essential for the economic development of an agriculture-based developing country. For this reason, we propose a method to detect flood by applying K-Means Clustering Algorithm and Color Probability techniques. The method performs background subtraction with the dynamic K-Means Clustering Algorithm to separate the background from the input flood image. Then connected component labelling will extract blobs. After that, morphological closing will fill small background color holes and color probability will find the watercolour pixels to detect the flooded region from the input flood image. Following this technique, we have achieved 96% accuracy.

14.1   INTRODUCTION

Bangladesh is vulnerable to floods [1] because of its geographical position. Due to heavy rainfall, when a flash flood occurs, sometimes brine enters our locality and damages crops severely, which harms our agricultural production as well as our economy. For example, in 2020, the floods damaged seedbeds of Aman rice worth Taka 15.95 crore and Aush rice worth Taka 49.79 crore in Rangpur, causing losses to seedlings. In total, floods damaged crops worth Taka 499 crore in Rangpur region in 2020. Therefore, it is very important to detect flooded regions to reduce flood-driven losses in agriculture. According to the 1988 Flood Archive records, more than 40% of the country as a whole been affected by the flood of 1987 [2]. There is, therefore, an urgent need to develop a system to identify possible flooded areas so that further precautions can be taken by Bangladesh's government and people living on suspicious land to deal with the disaster. To minimize the risk of flooding, the proposed method would raise awareness among people so that they can receive information and take necessary emergency measures during floods. Our main aim is to analyze previous flood records using a clustering method and to spot areas in Bangladesh that are vulnerable to flooding. Monitoring activity cannot be a priority for some individuals that are engaged in their day-to-day work [3].To minimize the risk of flooding, several flood risk technologies have been developed over the last few decades. Most of these technologies are designed for use in weather forecasting, flood detection, and monitoring systems using sensing devices, modeling software, the Internet, and mobile technology [4], which are generally for one-way communication only. Local communities need to access the website to get the most up-to-date information. To access the website, one needs a computer or a smartphone with an Internet service, which many people cannot afford to purchase. In the last few years, there have been quite a few dangerous flash floods all over the world. Many people died and lost their homes. One way to alert people regarding this type of disaster is by early detection of floods, which in turn reduces financial losses. Figure 14.1 shows how the flooded region can be detected from the input image.

In this chapter, we propose a rapid flood detection approach based on color probability, which will be helpful for the early detection of floods in Bangladesh. The key contributions of our proposed method can be summed up as follows:

Figure 14.1 Flooded region detection from the input image.

- Effectively detecting flooded areas from the input image

- Rapid flood-prone region detection

- Improved computational complexity, performance, and accuracy than that of the existing approaches

The rest of the chapter is organized in the following sections. Section 14.2 provides the literature review related to this arena. Section 14.3 discusses the proposed methodology. Section 14.4 analyses the result and finally, with section 14.5, the chapter is concluded.

14.2 LITERATURE REVIEW

Efficient flood detection has proven to be an important research topic in recent years. Most of the existing flood detection methods are based on satellite or aerial images [5]. This is not a trivial task to retrieve. Recently, many researchers have been working with flood detection and are proposing new methods for flood detection. A motion detection method is proposed in [6], which is based on background subtraction and spatial color information where fuzzy theory for background update is complex. A method based on flash floods is proposed in [7] for flood detection. However, in this method, the calculation of the boundary is time-consuming. A method based on the canny edge operator is proposed in [8] that works based on a threshold. In [9], the authors suggested a riveting approach for the detection of the flood without background subtraction, which is a probabilistic paradigm for flood detection in video sequences. However, in certain cases, it may be difficult to extract flood features and measure the spatial distribution. Moreover, in [10], the authors have proposed an early fire and flood detection method based on the integration of color, shape, and spatial analysis. The automatic calculation of the threshold used in this approach is not a very effective one. In [11], the authors have proposed an almost real-time approach for flood detection. However, it only achieves instantaneous calculation for images having low resolution. High quality images require extra processing time. A comparison of three models, named Evidence-based Belief Function (EBF), Random Forest (RF), and Boosted Regression Trees (BRT) is provided in [12] to classify flood-prone areas in the Galikesh region of Iran. EBF with an accuracy of 78.67% is the best among these three techniques in terms of altitude, slope, and the topographic wetness index. Though EBF produced comparatively better output, it requires more computational time as it requires reclassification of conditioning factors. Moreover, RF has the ability to process different scales of measurement with the fastest time among these three, although it has the lowest level of accuracy (73.33%). The authors of [13] produced a map showing the flood affected zone with the kernel K-means approach in the southwestern part of Madagascar. Non-linear clustering kernel k-means with log-ratio images have been suggested to address the challenges raised by floods. However, this method has a limitation of obtaining images in real time during floods. By considering various factors as attributes such as precipitation, climate change, researchers have tried to figure out the most spatial features that cause flooding in North Texas using Bayesian algorithms in [14].

The methods discussed above used feature extraction for the processing of flood images. Meanwhile, convolutional neural networks also gained some success. In [15], the FCN AlexNet deep learning model is used to detect floods in susceptible areas. However, it may not be implemented in complex enviornment as the number of classification classes is only three. The authors of [16] uses the Random Under-Sampling Boosted (RUSBoost) classification algorithm on Cyclone Global Navigation Satellite System (CYGNSS) data. However, it suffers from overestimation of flooded areas for images of different dataset other than CYGNSS. In [17], the authors proposed a method based on the deep convolutional neural network to identify flooded region. Though the technique produced an approximately instant output, the flood mapping obtained in this method is not very effective. The authors of [18] suggested a semantic segmentation model U-Net to track landmass. The outcome of the model is further checked with flood sensors for validating the result. However, it may not predict the segmentation mass correctly if the landmass is irregular. Though various deep learning techniques are being used nowadays, they are not effective for concurrent flood detection. This is because they require higher computational time than feature extraction techniques.

Effective measures cannot be taken at the earliest possible time in the context of Bangladesh by using the above-mentioned approaches. As a result, in our country, early detection of a flood is a challenging task. Thus, we propose a rapid flood detection approach based on color probability that will help in the early detection of floods in Bangladesh.

14.3   OUTLINE OF METHODOLOGY

The necessary steps of our proposed method for flood detection are shown in Figure 14.2, which are explained in the following sub-sections. The first step of the proposed method is background subtraction to distinguish the foreground from the input image background. The next step is connected component labeling, which will extract blobs. After applying morphological closing and color probability, small background color holes will be filled and the watercolor pixels will be found. Finally, through performing edge analysis, a final decision is made as to whether the region is flooded or not.

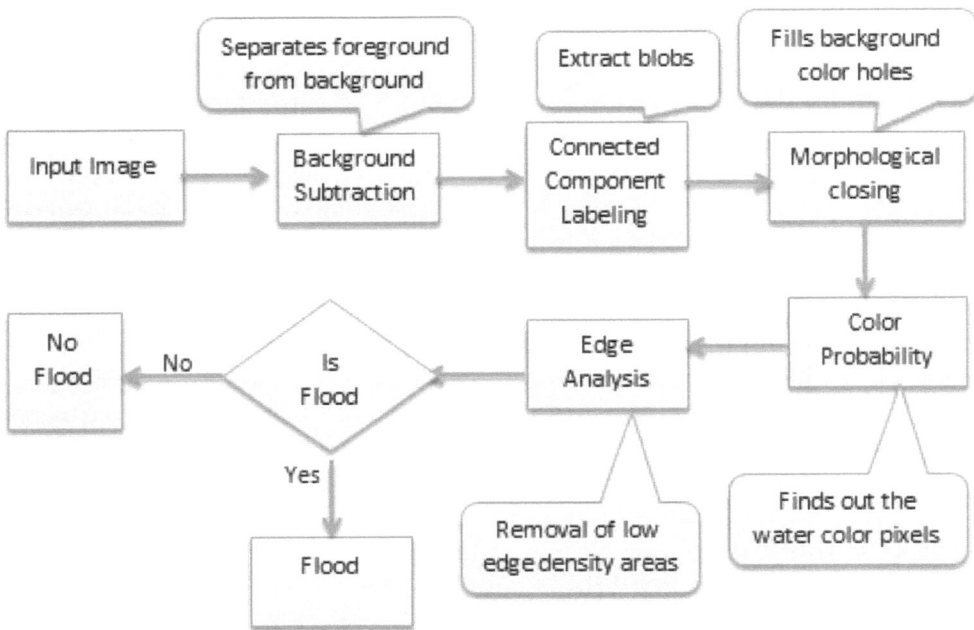

Figure 14.2   The proposed framework for detecting the flooded region.

14.3.1   Background Subtraction

Background Subtraction is a method that aims to isolate a relatively stationary background from foreground objects. The reason for choosing background subtraction is that it can detect any changes that occur in a scene. Typically, the flood comprises not only large amounts of water but also soil and objects, such as pieces of benches, buildings, and so on. Non-water additions are mixed with water and make a distinction. Flood surfaces may be either brownish or greyish [19]. The relationship between red (R), green (G), and blue (B) channels can characterize the brownish part of the flood. Saturation (S) tends to be poor in rare instances where water has a grey appearance. The flood will be measured by its color attribute by incorporating the facts referred to above, and the mean and variance values for the R, G, B, and S channels will be calculated. In detail, the core of the background subtraction process is shown in Figure 14.3.

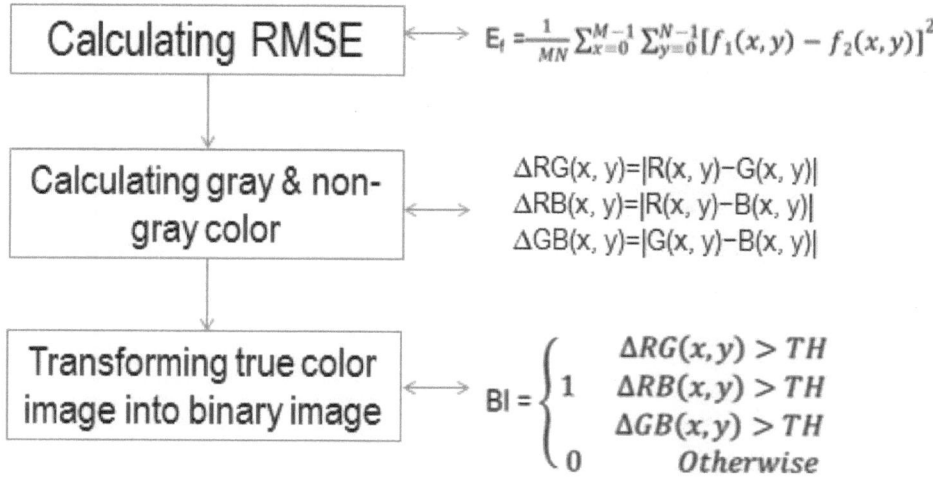

Figure 14.3 Background subtraction procedure.

14.3.2 Dynamic K-Means Clustering Algorithm

Based on the color of the flash flood, the pixels need to be clustered together. Here, we propose an K-Means clustering algorithm, which is referred to as the dynamic K-Means clustering algorithm. Without specifying the number of clusters (K) value, the proposed method would dynamically cluster all data from a broad data set. Choosing the initial cluster value (K) is a very difficult task. That is why we have proposed an automatic clustering technique that strengthens the outcome of clustering. The flood image, shown in Figure 14.4 within the sample dataset is solved by the dynamic K-Means approach for the research.

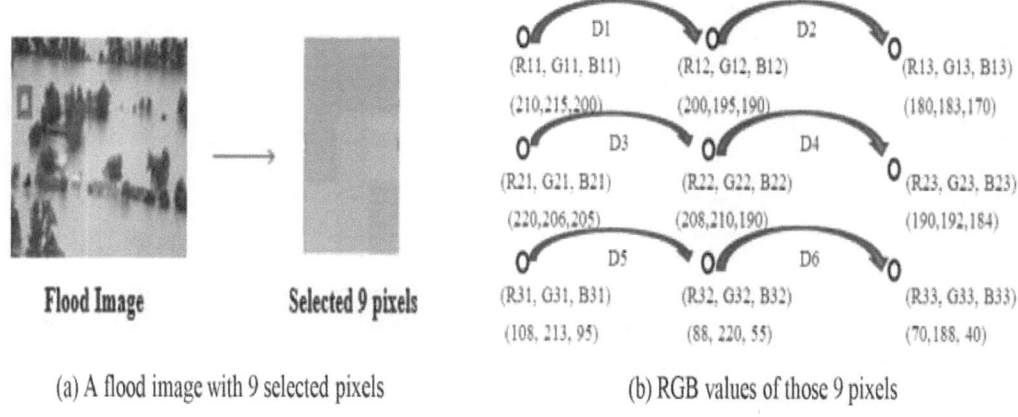

(a) A flood image with 9 selected pixels   (b) RGB values of those 9 pixels

Figure 14.4 Dynamic K-Means procedure for some selected pixels in an input image.

The Euclidean distances of these 9 pixels based on their RGB values are D1 = 17.32, D2 = 30.72, D3 = 19.62, D4 = 26.15, D5 = 45.27 and D6 = 39.66. From these distances, we get the distance matrix, which is depicted in Figure 14.5.

	D1	D2	D3	D4	D5	D6	MEAN,μ
D1	0	13.4	2.30	8.83	27.95	22.34	μD1=12.47
D2	13.4	0	11.10	4.57	14.55	8.94	μD2=8.76
D3	2.30	11.10	0	6.53	25.65	20.04	μD3=10.94
D4	8.83	4.57	6.53	0	19.12	13.51	μD4=8.75
D5	27.95	14.55	25.65	19.12	0	5.61	μD5=15.48
D6	22.34	8.94	20.04	13.51	5.61	0	μD6=11.74
							μD=Th=13.36

Figure 14.5 Distance Matrix.

The initial centroid is D4. If the threshold is greater than or equal to the distance between the centroid and the data point, then they will be grouped in the same minimum distance group; otherwise, they will be grouped in another group. Using this concept, we have two data groups. Therefore, the number of clusters in the first iteration is two. In this way, as long as the cluster value changes, the above procedure iterates. Figure 14.6 shows the initial two data groups from the first iteration and their corresponding centroids.

Data group 1	Data group 2
D1=17.32	D5=45.27
D4=26.15	D6=39.66
D2=30.72	Centroid, C5= 42.47
D3=19.62	
Centroid, C3=23.45	

Figure 14.6 Initial data groups and their respective centroids.

At the first iteration, following our proposed dynamic K-means clustering algorithm, we have 2 clusters. However, in the traditional K-means approach, at the initial stage, the clusters are selected randomly and then the optimal positions of the centroids are iteratively calculated. If the traditional clustering algorithm was to apply on the above 9 pixels from Figure 14.4(a), after the first iteration the number of clusters would be 3. This K-means approach takes more time to get the optimal positions, which is because it requires more iterations, whereas in our proposed method, we have calculated the K value dynamically, which reduced the computational time to a great extent. The working procedure of the proposed dynamic K-Means clustering algorithm is shown in Figure 14.7.

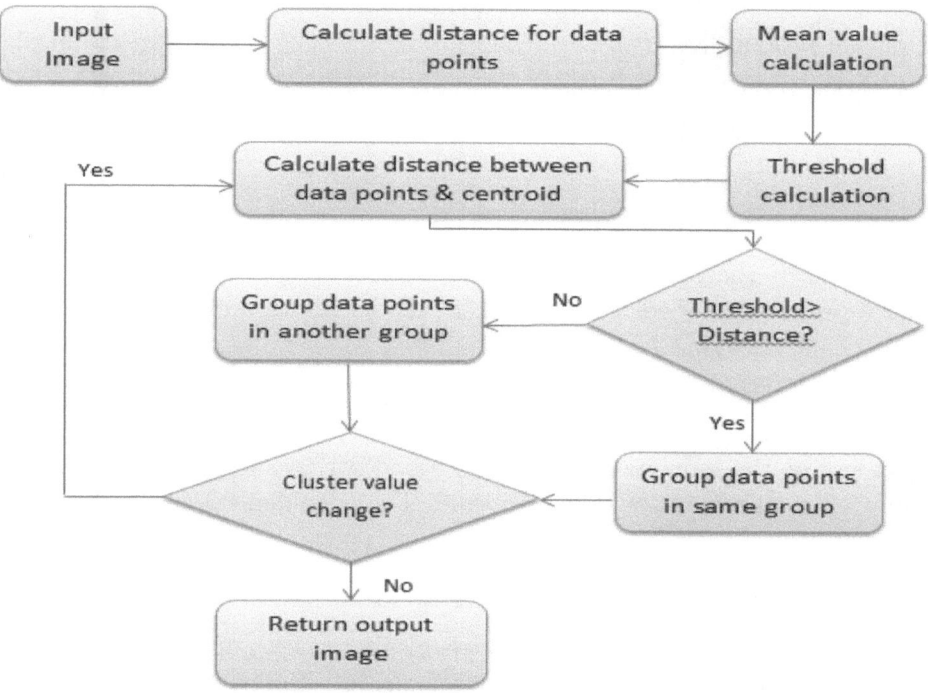

Figure 14.7 Working principle of the dynamic K-Means clustering algorithm for flood region detection.

14.3.3 Connected Component Labelling

An 8-directional connected-component labelling algorithm joins the pixels in the next processing stage. Connected component labelling works to identify connected pixel regions, i.e. neighboring pixel regions that share the same intensity value set, V, by scanning an image, pixel-by-pixel (from top to bottom and from left to right).

14.3.4   Morphological Closing

Morphological closing may unite the separated pixels into a morphological closure [20]; [21]. Dilation and erosion are two basic binary mathematical morphologies. Erosion eliminates small-scale information from a binary image but decreases the size of the regions of interest at the same time. Dilation is the opposite of erosion.

In mathematical morphology, the closing of a binary image A by a structuring element B is the erosion of the dilation of that set.

$$A \cdot B = (A \oplus B) \ominus B \tag{14.1}$$

where $\oplus$ and $\ominus$ denote the dilation and erosion operation, respectively.

14.3.5   Color Probability

Color probability refers to the disability to distinguish colors. The authors in [8] can determine the normalized color probability of a pixel that belongs to the flood area –

$$C_i(x, y) = e^{-\frac{(E_i(x,y)-\mu_i)^2}{2\sigma_i^2}} \tag{14.2}$$

where i is one of the four channels of the current E pixel (x, y). $C_i(x, y)$ is the normalized probability of color according to the distribution of the ith channel.

14.3.6   Edge Density

The edge density is another means of distinguishing floods and other objects. A moving object can be the same color as a flood. Flood surfaces usually have many edges, unlike greyish road regions or brownish sand. The edge density can be calculated in the following way.

$$D_E = \frac{A_e}{A_b} \tag{14.3}$$

where A_e refers to the number of edge pixels inside the blob and the total number of pixels in the blob is A_b. Low edge density regions are eliminated.

Finally, we get the desired output from the sample input flood image as shown in Figure 14.8. The red colored pixels refer to the flooded region of the input image.

14.4   EXPERIMENTAL RESULT ANALYSIS

To assess the performance of the proposed algorithm and whether the results are obtained as per the objectives, validation of the algorithm on several datasets is necessary. The proposed approach is evaluated over a dataset containing approximately 300 images. The dataset was prepared by the authors by collecting a few images from the Department of Water Resource Engineering (WRE), Chittagong University of Engineering & Technology (CUET) and some other personal sources. The dataset is comprised of flooded and non-flooded images. The non-flooded images consist of images with different lakes, ponds, rivers, seas, etc. Some sample images from our dataset are shown in Figure 14.9.

(a) Flood Image   (b) Output Image

Figure 14.8 Resultant flood detected region using our proposed method.

Figure 14.9 Sample images from our dataset.

Intel Core i7 870 CPU working at 2.93GHz and 8GB DDR3 RAM has been used for the evaluation process. The system has been executed in an experimental platform built in JAVA. Our proposed algorithm has achieved accuracy of nearly 96%. Figure 14.10 shows some sample input flood images and their corresponding flood-detected region as output using our proposed method.

Table 14.1 shows that the method presented in this chapter has achieved real-time efficiency in terms of TPR, FPR, TNR, Recall, Precision, and F1 score for images shown in Figure 14.10.

By plotting the True Positive Rate (TPR) and False Positive Rate (FPR) on a graph, the Region of Convergence (ROC) curve can be acquired. For the given

(a) Sample Images (b) Flood detected areas

Figure 14.10 Applying our proposed algorithm on sample input images from our dataset.

Table 14.1 Accuracy measurement of samples for the proposed method.

Sample	TP	FP	TN	FN	TPR	FPR	Recall	Precision	F1 score
Sample 1	30153	218	2653	347	98.86	7.59	98.86	9.28	99.07
Sample 2	8961	235	233	1691	97.47	12.2	97.47	97.44	97.45
Sample 3	4391	39	599	1344	87.996	20.14	87.996	92.83	90.34

samples in Figure 14.10, the ROC curve can be obtained by taking TPR along the Y-axis and FPR along the X-axis as shown in Figure 14.11.

We have compared our proposed method with the recent state-of-the-art techniques [22]. Figure 14.12 exhibits the graphical comparison between [22] and our proposed technique. It shows that our proposed method can detect the flooded region more accurately and precisely.

We have compared our proposed algorithm with the above stated algorithms in terms of various accuracy measures, which is shown in Table 14.2. From the table, it is evident that our method gives better results than other recent approaches in this arena. All the methods are simulated by the authors using the images from our dataset.

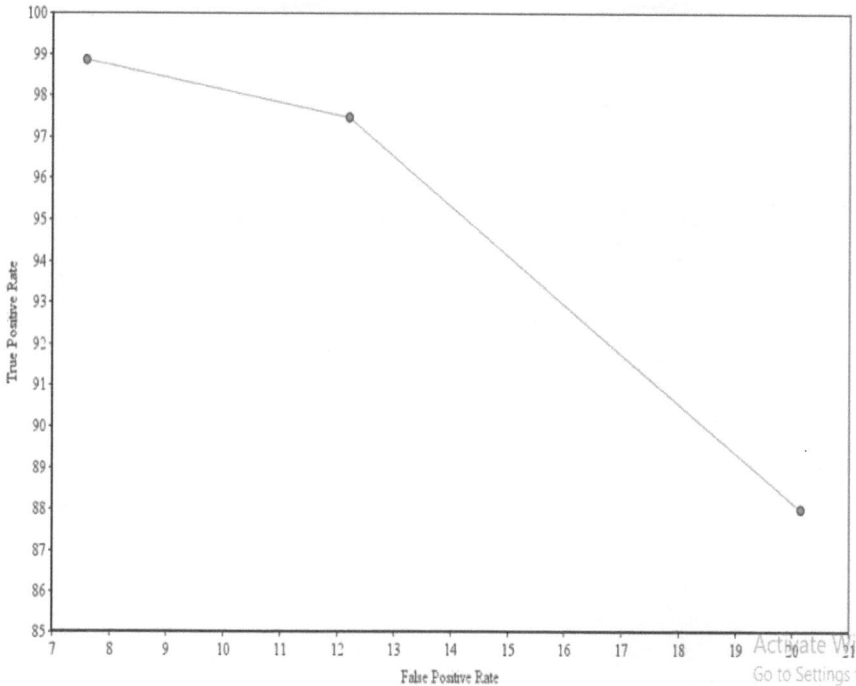

Figure 14.11 ROC Curve.

Table 14.2 Comparison among recent methods of flood region detection and the proposed method.

Accuracy measure	Kernel K-Means method	EM method	Hierarchical method	Proposed method
F1 score	94.51	94.33	91.21	**95.62**
TP rate	93.68	94.48	90.29	**94.78**
FP rate	13.46	14.44	17.49	**13.31**
Precision	95.37	94.18	92.19	**96.52**

Bold data represents the best result among the given data.

As proposed by [22], the best method among Kernel K-means, Expectation Maximization (EM) and Agglomerative Hierarchical Clustering is Kernel K-means clustering model, because it provides better accuracy than the other two methods. A comparison of accuracy between the best method of [22] and our proposed method is shown in Table 14.3.

We have also evaluated our method in terms of computational complexity with the Kernel k-means clustering technique of [22]. Table 14.4 depicts that our method requires comparatively less time than the other technique. Though Kernel K-means

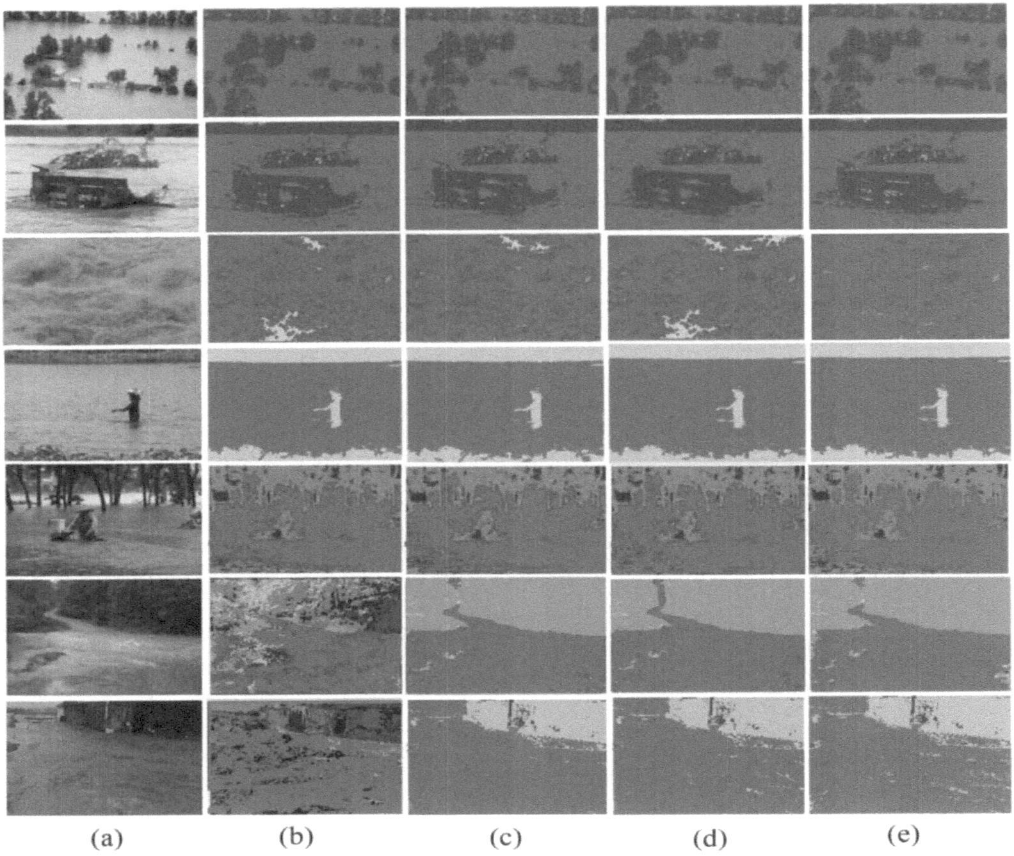

Figure 14.12 Comparison among different flood region detection technique. (a) flooded image, (b) Kernel K-means method, (c) EM method, (d) Hierarchical method, (d) proposed method.

Table 14.3 Comparison between the best method of [22] and the proposed method.

Reference name	F1 score	TP rate	FP rate	Precision
[22]	94.51	93.68	13.46	95.37
Proposed method	**95.62**	**94.78**	**13.31**	**96.52**

Bold data represents the best result among the given data.

clustering technique offers less time for small images, its computational time increases as the size of the image increases. The methods are evaluated under similar software and hardware condition.

Figure 14.13 depicts how accurately our method can distinguish between flooded and non-flooded images. Our approach can precisely label different water bodies such as lakes, ponds, rivers, etc. as non-flooded images, whereas other techniques have

Table 14.4 Comparison between 3 means clustering technique and our proposed method in terms of computational complexity.

Image Size	Kernel K-Means method (ms)	Proposed method (ms)
390×263	**1172**	1281
400×300	1187	**1047**
600×500	2782	**2719**
700×600	2297	**2282**
900×800	4406	**4265**

Bold data represents the best result among the given data.

shown some anomalies in the resultant images while evaluating non-flooded images. The red pixels in the resultant images of Figure 14.13 represent the flooded region.

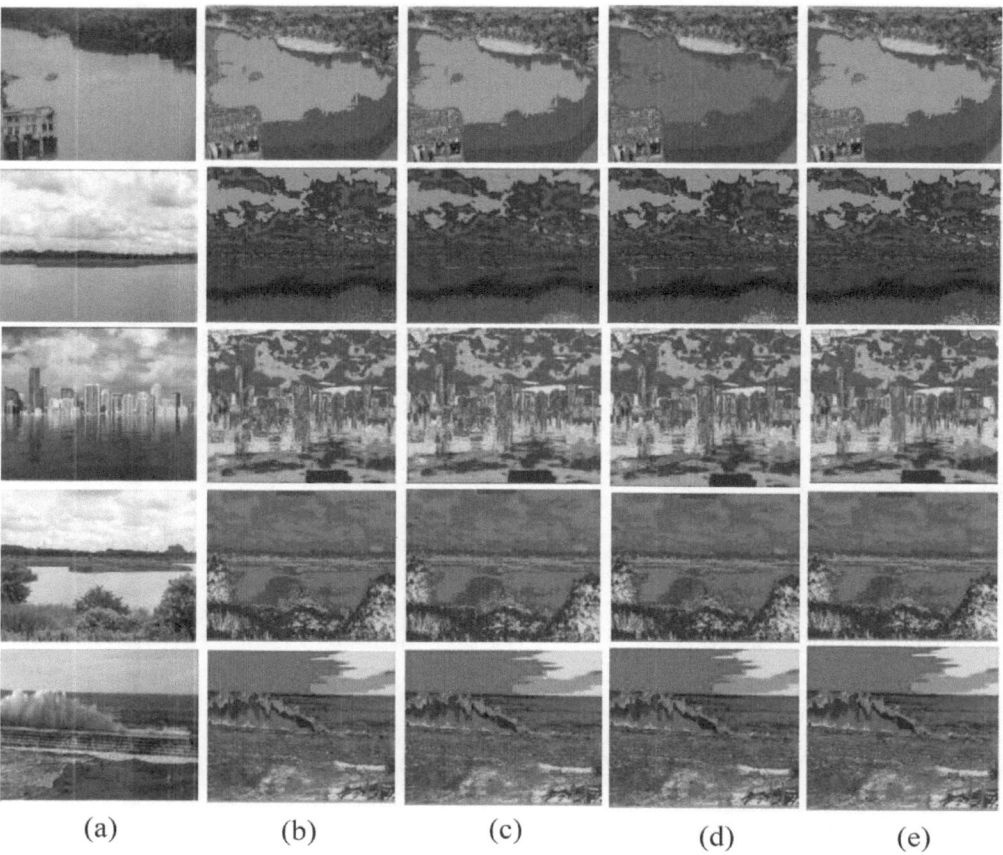

(a) (b) (c) (d) (e)

Figure 14.13 Comparison among different flood detection methods for non-flooded images. (a) flooded image, (b) Kernel K-means method, (c) EM method, (d) Hierarchical method, (d) proposed method.

In our proposed algorithm, we performed background subtraction in the input image to separate the foreground of the image. Then we performed Dynamic K-Means clustering on the pixels based on their color. Here, we introduced an automatic means of generating the K value, which makes our approach computationally faster. After that, connected component labelling was used to extract blobs. Through color probability, the water color pixels are found and then components with lower edges are discarded as water bodies generally have the high number of edges. Our technique achieves better accuracy than recent state-of-the-art methods in detecting flooded areas. It can also precisely distinguish between flooded and non-flooded regions.

14.5 CONCLUSION AND FUTURE WORK

A fast flood region detection approach based on the dynamic K-Means algorithm and color probability is presented in this chapter. The processing speed is fast enough for surveillance cameras which are being used nowadays. Here, we used background subtraction to acquire the foreground from the input flood image. After that, we employed connected component labeling to extract blobs and morphological closing to fill the blobs with tiny background color holes. Finally, the color analysis determines the pixels that are of watercolor and edge analysis to eliminate low edge density regions. The proposed approach is capable of detecting floods and is sufficiently accurate. For moving cameras, the flood detection algorithm can be expanded in the future.

ACKNOWLEDGEMENT(S)

The authors sincerely thank the Department of Water Resource Engineering (WRE), Chittagong University of Engineering & Technology (CUET) for their utmost support in the formation of the dataset by providing the flooded area images of Bangladesh.

Bibliography

[1] Geography of Bangladesh. (2017). Retrieved from https://en.wikipedia.org/wiki/Geography_of _Bangladesh.

[2] Greenfieldboyce, N. (2007). Study: 634 Million People at Risk from Rising Seas. *Napier*. Retrieved from https://www.npr.org/templates/story/story.php?storyId=9162438.

[3] Pagatpat J. C., Arellano A. C. and Gerasta O. J. (2015). GSM & web-based flood monitoring system. *IOP Conference Series: Materials Science and Engineering*, 79.

[4] Vitales J. S., Villajin L. D., Destreza F. G., Ricafranca D. V., and Rodriguez. (2016). Flood detection and monitoring system using modeling software, sensing devices, internet and mobile technology. V. M. R. 2016 College of Engineering and Computing Sciences, (Batangas, Philippines: Batangas State University ARASOF Nasugbu).

[5] Giustarini L., Hostache R., Matgen P., Schumann G. J. -P., Bates P., D., Mason D. C. (2013). A change detection approach to flood mapping in urban areas using TerraSAR-X. *IEEE Transactions on Geoscience and Remote Sensing, 51*(4), 2417–2430.

[6] Harrouss O. E., Moujahid D., Tairi H. (2015). Motion detection based on the combining of the background subtraction and spatial color information. *2015 Intelligent Systems and Computer Vision(ISCV).* (pp. 1-4).

[7] Filonenko A., Wahyono, Hernández D. C., Seo D., Jo K. (2015). Real-time flood detection for video surveillance. *IECON 2015 - 41st Annual Conference of the IEEE Industrial Electronics Society.* (pp. 004082-004085).

[8] Dong Y., Li M., Li J. (2013). Image retrieval based on improved canny edge detection algorithm. In *Proceedings of 2013 International Conference on Mechatronic Sciences, Electric Engineering and Computer (MEC).* (pp. 1453-1457).

[9] Borges P. V. K., Mayer J., Izquierdo E. (2008). A probabilistic model for flood detection in video sequences. In *15th IEEE International Conference on Image Processing.* (pp. 13-16).

[10] Lai C. L., Yang J. C., Chen Y. H. (2007). A real time video processing based surveillance system for early fire and flood detection. In *Proceedings of 2007 IEEE Instrumentation & Measurement Technology Conference IMTC 2007.* (pp. 1-6).

[11] Mason D. C., Davenport I. J., Neal J. C., Schumann G. J. -P., Bates P. D. (2012). Near real-time flood detection in urban and rural areas using high-resolution synthetic aperture radar images. *IEEE Transactions on Geoscience and Remote Sensing, 50*(8), 3041–3052.

[12] Rahmati O., Pourghasemi H. R. (2017). Identification of critical flood prone areas in data-scarce and ungauged regions: A Comparison of Three Data Mining Models. *Water Resources Management, 31*, 1473—1487.

[13] Razafipahatelo D., Rakotoniaina S., Rakotondraompiana S. (2014). Automatic floods detection with a kernel k-means approach. In *Proceedings of 2014 IEEE Canada International Humanitarian Technology Conference - (IHTC).* (pp. 1-4).

[14] Jangyodsuk P., Seo D., Elmasri R., Gao J. (2015). Flood prediction and mining influential spatial features on future flood with causal discovery. In *Proceedings of 2015 IEEE International Conference on Data Mining Workshop (ICDMW).* (pp. 1462-1469).

[15] Son K., Yildirim M. E., Park J., Song J. (2019). Flood detection by using FCN-AlexNet. In *Proceedings Volume 11041, Eleventh International Conference on Machine Vision (ICMV 2018).* 110412P.

[16] Ghasemigoudarzi P., Huang W., Silva O. D., Yan Q., Power D. T. (2020). Flash flood detection from CYGNSS data using the RUSBoost algorithm. *IEEE Access*, *8*, 171864–171881.

[17] Jain P., Schoen-Phelan B., Ross R. (2020). Automatic flood detection in Sentinel-2 images using deep convolutional neural networks. In *Proceedings of the 35th Annual ACM Symposium on Applied Computing.* (pp. 617–623).

[18] Basnyat B., Roy N., Gangopadhyay A. (2021). Flood detection using semantic segmentation and multimodal data fusion. In *Proceedings of 2021 IEEE International Conference on Pervasive Computing and Communications Workshops and other Affiliated Events (PerCom Workshops).* (pp. 135-140).

[19] Soille P. (1999). Morphological image analysis: principles and applications. *Berlin: Springer, 2*(3).

[20] Lai C. L., Yang J. C., Chen Y. H. (2007). A real time video processing based surveillance system for early fire and flood detection. In *Proceedings of 2007 IEEE Instrumentation & Measurement Technology Conference IMTC 2007.* (pp. 1-6).

[21] Borges P. V. K., Izquierdo E. (2010). A probabilistic approach for vision-based fire detection in videos. *IEEE Transactions on Circuits and Systems for Video Technology, 20*(5), 721–731.

[22] Raihana K. K., Rishad S. M. K., Sadia T., Ahmed S., Alam M. S., Rahman R. M. (2018). Identifying flood prone regions in Bangladesh by clustering. In *Proceedings of 2018 IEEE/ACIS 17th International Conference on Computer and Information Science (ICIS).* (pp. 556-561).

CHAPTER 15

Fabrication of Smart Eye Controlled Wheelchair for Disabled Person

Md. Anisur Rahman

Chittagong University of Engineering and Technology, Chattogram, Bangladesh

Md. Abdur Rahman

Chittagong University of Engineering and Technology, Chattogram, Bangladesh

Md. Imteaz Ahmed

Chittagong University of Engineering and Technology, Chattogram, Bangladesh

Md. Iftekher Hossain

Chittagong University of Engineering and Technology, Chattogram, Bangladesh

CONTENTS

THIS work focuses on fabricating an innovative wheelchair prototype controlled through eye movement for disabled persons. Nowadays, a large number of people suffer from various disabilities, including hand and leg disabilities. Since disability causes many problems in afflicted people's regular life, this issue must be checked to improve their lifestyle. A paradigm for controlling a wheelchair by eye movement detection is proposed in this study. This study aims to develop a feasible paradigm for helping disabled persons with their everyday needs. The wheelchair is controlled with the assistance of a smart glass. A camera is mounted on the glass, capturing the eye pupil's image without hindering eyesight. An Image processing technique has been used to detect eye movement. There is also an interface for calibrating light in different environments. After proper assembly, the prototype has been controlled easily through eye movement.

15.1 INTRODUCTION

About 15 percent of the world's population suffer from some form of disability [1]. Some of these people have neuromuscular disorders such as cerebral palsy, spinal cord injury, brainstem stroke, amyotrophic lateral sclerosis etc. In those disorders, people can hardly use their muscles. As a result, their daily life activity is hampered greatly. They can't even move from one place to another place without the help of others. Some technological solutions must be introduced to make their life better. Researchers have been trying to develop new technologies to help them control their wheelchairs. Some of the recent techniques are voice, head stick, tongue switch, etc. [14] [13]. Head movements or a head-mounted accelerometer or inclinometer, brain waves (EEG) and muscles on the face (EMG) are all potential input sources [16]. The benefits of an eye tracking-based device over the other alternatives are numerous. In a voice-controlled wheelchair surrounding noise can create a problem. In a head movement-based system, the wrong movement creates a problem. As a result, we centered our efforts on developing a human-computer interface based on eye movement that analyzes and recognizes the eye movement patterns and uses the patterns to operate a wheelchair. Disabled people use a lot of their energy for turning and moving the wheelchair. Those people would conserve their energy and use their hands and arms for other tasks if they use their eyesight as their guide. The focus of this project is to create a wheelchair that can be operated by using the eyes of a disabled person. Although the infrared reflected eye detection system provides accurate results, infrared radiation is harmful to the eyes [17]. Some researchers have used different algorithms in camera mounted eye detection techniques, such as the Kalman filter, Hough transform, Haar cascade, etc. [2][1][15]. Some techniques requires face landmark recognition [7]. However, these algorithms and techniques are computationally expensive and take some time for processing images and executing the command. So, in this chapter, a novel eye movement control technique is introduced, which is safe, easy to use and fast enough for real-time application. In addition, it is lower in than most of the recent eye detection devices. In the early days, for disabled or novice people, using a computer was considered a tedious task. This is why introducing eye movement detection as an input system for machines/computers can provide benefits for disabled people.

15.2 RELATED WORK

In the recent century, there has been an increasing number of published articles on Human-Computer Interaction (HCI). Most researchers gave importance to designing the flexible interface and maintaining ergonomics issues. The first research was a non-invasive eye-tracking technique. This noninvasive eye tracking was done by using corneal light reflection [6]. In this system, the participant's head needs to be motionless, and the system recorded only horizontal eye position. In order to record the temporal aspects of eye movements in two dimensions, motion picture photography was applied in 1905. In this motion picture photography, the researchers recorded the movement of a small white speck of material placed on the participants' corneas rather than light reflected from the cornea directly. The first head-mounted eye tracker was designed in 1948 [8]. Later on, in 1974, two combined military teams created an eye-tracking system that significantly reduced tracker obtrusiveness and participant limitations [12]. Most of the published research is on psychological theory and eye-tracking technology, linking eye-tracking data to cognitive processes, military aviation research, punched paper cards and tapes, teletypes for command line entry, and printed lines of alphanumeric output served as the primary research of human computer interaction. The inception of real-time eye tracking was in the 1980s. The early research in real-time eye tracking focused on disabled users. Researchers began to look into how eye tracking could be used to help with human-computer interaction concerns [3]. In short, researchers were primarily concerned with making the interaction between humans and machines as flexible as possible [9].

15.3   NOVELTY AND CONTRIBUTION

In contrast to previous research, this project focused our efforts on designing a human-computer interface-based wheelchair. The novel contribution of the proposed project are as follows:

- Capture the patterns of real-time eye movement, then analyze and process the data to recognize the users.

- Design a robust approach to control the movement and speed of the wheelchair according to the user's direction.

- A prototype of the eye pupil-controlled wheelchair is designed to assist the movement of the disabled person, who is unable to move their limbs.

15.4   ORGANIZATION OF THE CHAPTER

The rest of the chapter is arranged as follows. Section 15.2 gives an idea of systems design, including software and hardware configuration, Section 15.3 provides an overview of the methodology and a detailed flow chart. Section 15.4 presents the result of the proposed project, followed by a conclusion in Section 15.5.

15.5   SYSTEM DESIGN

15.5.1   Hardware configuration

The hardware architecture of the system is shown in Figure 15.1. Here Raspberry pi is the central processing unit in this system. It is a single-board computer. We have used the Raspberry pi 3B model for real-time data acquisition and analysis. Here Raspbian operating system is used. Smart glass is used in this system shown in fig. 15.2. The Raspberry pi Camera is attached inside the glass. The camera is attached at such an angle that when we look at the left side, the entire pupil portion can be captured in the camera; when we look forward, some pupil portion can be found, and when we look at the right side, no pupil portion can be found. This particular setup gives the advantage of differentiating three different eye movements.

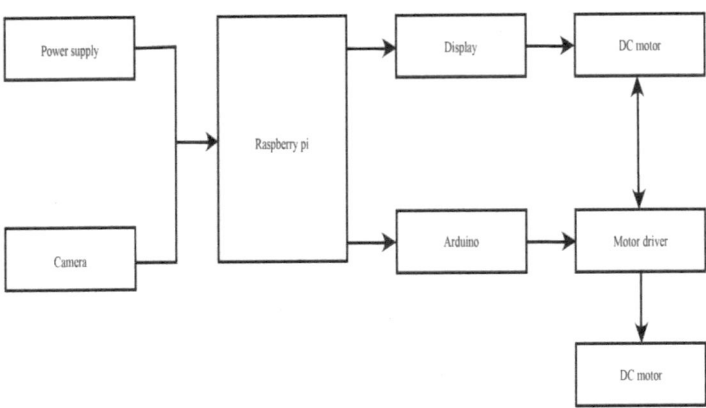

Figure 15.1 Hardware architecture.

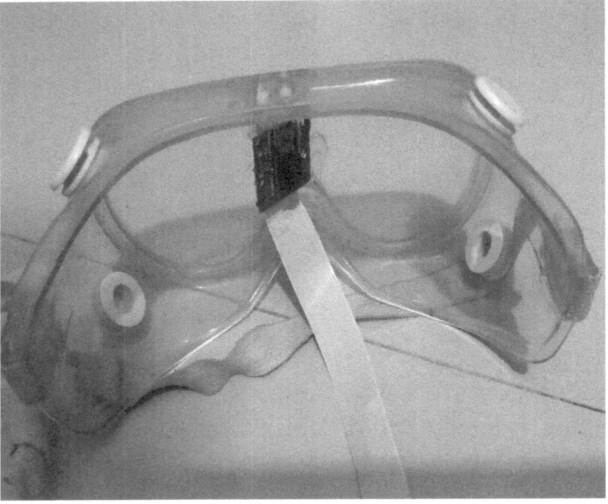

Figure 15.2 Smart glass.

A prototype of the wheelchair is shown in Figure 15.3. It can move along in a right-left or forward-backwards direction according to the controller's instructions. Arduino is a single board micro-controller for building digital devices. In this system,

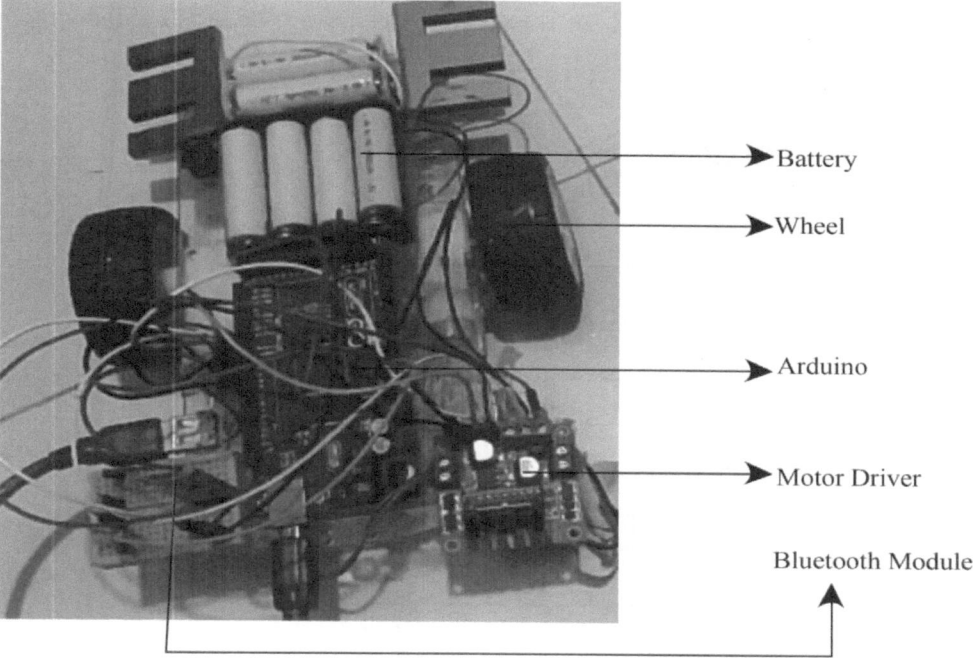

Figure 15.3 Wheelchair prototype.

Arduino Uno is used to control the demo wheelchair. There is a Bluetooth module for Raspberry pi-to-Arduino serial communication. Here an HC06 Bluetooth module is used for the system. The motor driver controls the motor direction. Here an L298 motor driver is used. Two 5-volt DC motors are used in the demo wheelchair. Six 1.5-volt rechargeable batteries are used as a power supply for the demo wheelchair.

15.5.2 Software configuration

In our system, the Python, open computer vision (CV) library is used for image processing, because the open CV library has different features which can help detect eye movement [4]. To find the pupil portion effectively, some steps have been followed:

1. BGR to Gray conversion: A color image is converted into a gray image by using this library. It is done to reduce the system delay time.

2. Blurring image: For blurring images, a Gaussian blur filter is used. This can help find the exact edges of a specific area. It also reduces the noise in the image.

3. Thresholding: Thresholding is an image segmenting method. Thresholding can be used to create a binary image from a grayscale image. Thresholding replaces each pixel in an image with a black pixel if the intensity of the image is less than some fixed constant, or a white pixel if the intensity of the image is greater

than the constant. From the thresholding image a total number of pixels, not black, can be determined. In this project, we have used this number to control the different movement of the wheelchair.

15.6 METHODOLOGY

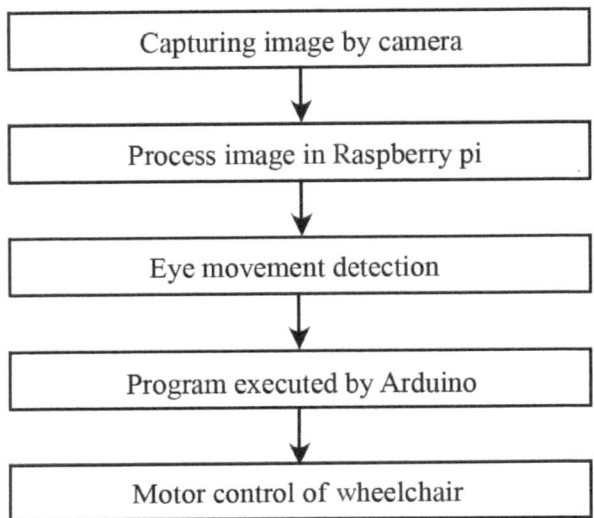

Figure 15.4 Functional block diagram.

The functional block diagram is shown in Figure 15.4. The camera captures the movement of the eye. Then it is processed and analyzed in Raspberry pi, and the processed eye movement information goes to Arduino via a Bluetooth module. Arduino sends the signal to the motor driver. The motor driver controls the direction of the motors, which are attached to the wheels. Thus, when a disabled person looks toward a direction, the wheelchair also moves towards the desired direction:

1. The Camera captures the position of the pupil.

2. Based on the pupil portion captured by the camera, the number of pixels not black or having a nonzero count in the thresholding image have been found. After analyzing the pixel values, two threshold values are developed.

3. Minimum and maximum threshold values are established.

If the pixel value is less than the maximum threshold value, the wheelchair will move towards the left side. Because the camera setup is inclined towards the left side gazing, the nonzero value will be higher. If the value is between the minimum and maximum threshold, then the wheelchair will move forward, and if the pixel value is less than the minimum value, then the wheelchair will move towards the right side. The minimum and maximum threshold values can vary based on the light intensities. In this experiment, the minimum threshold value was 1000, and the maximum threshold value was 6000.

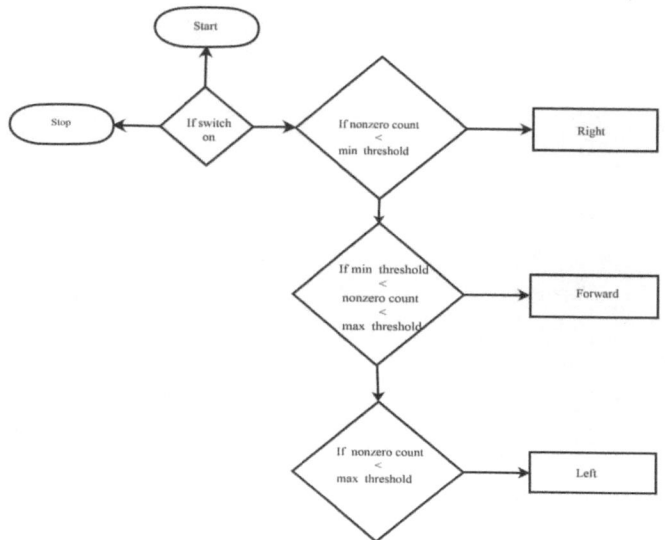

Figure 15.5 System workflow.

15.7   RESULT

The resulting image processing resulted data was sent to the wheelchair system, which was focused on the position of the eye pupil center. The command is sent to the motor driving circuit. The wheelchair then moves in desired direction based on eye movement. The result can be demonstrated as follows:

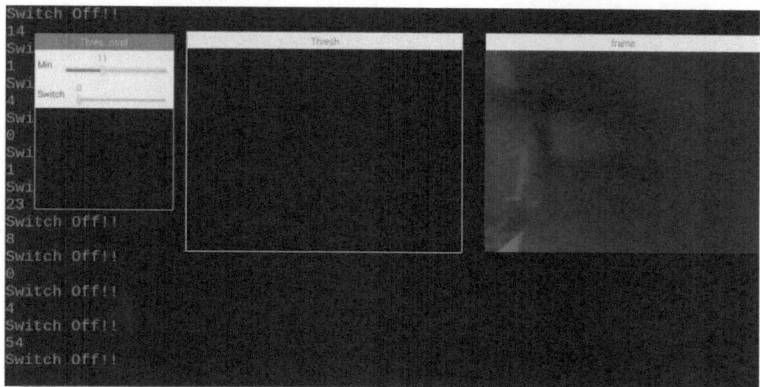

Figure 15.6 Eyes at the right position

The difference can be seen by looking at the threshold image. Here in Figure 15.6, almost no pupil segment is found in the camera in the right eye position. Therefore, there is no brighter section in the thresholding image. As a result, the nonzero count value is less than the minimum threshold value.

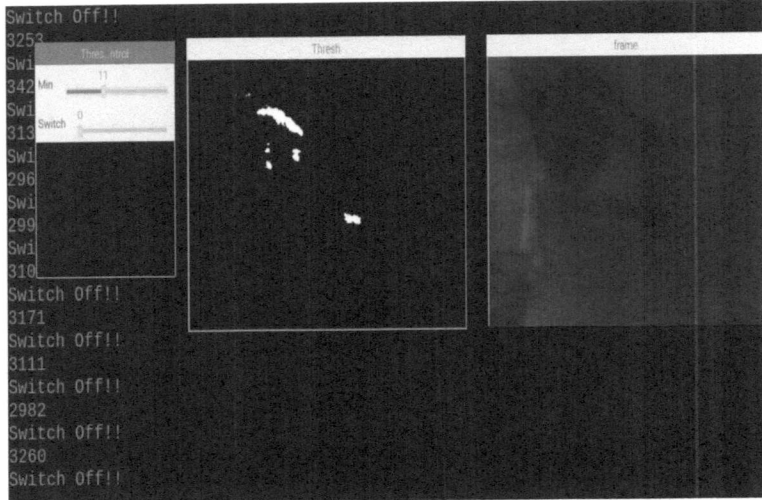

Figure 15.7 Eyes at the forward position

Nevertheless, in Figure 15.7, there is some pupil portion found in the camera in the forward position. Due to this, some white portion is in the thresholding image, so the total nonzero count is medium.

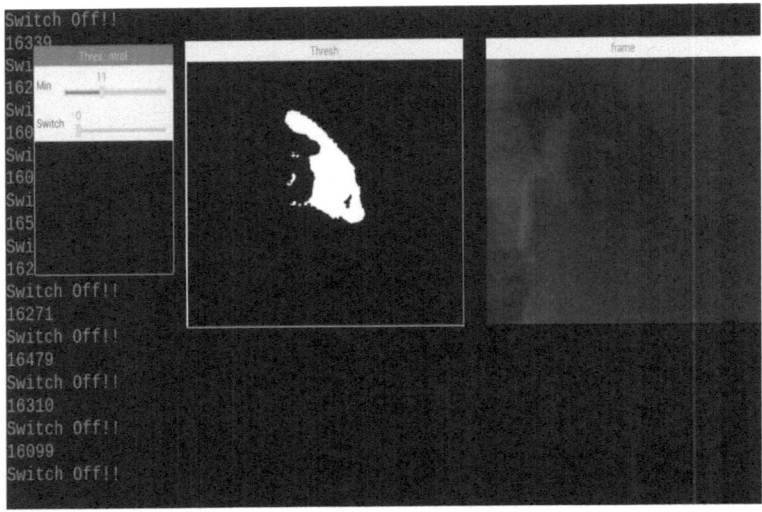

Figure 15.8 Eyes at the left position

In Figure 15.8, when we look at the left side, most of the white portion can be found in the threshold image as most pupil parts can be visible here. This is why, the total nonzero count is higher than the maximum threshold. For this difference in the visibility of images, creates the variation in threshold and the nonzero count. Using these 3 different value ranges, three positions can be divided, thus controlling the wheelchair robot to move to the desired direction. The wheelchair can be started or stopped by the digital switch; there is also a button attached in the demo wheelchair that serves the same purpose. A person who is partially disabled can use their legs or hand to start or stop the wheelchair. In the interface, there is another feature for calibrating light of different intensities. This feature makes the wheelchair movement more accurate.

15.8 CONCLUSION AND FUTURE WORK

Eye movement research to be a promising area of the applied field. This technique of eye movement detection effectively controls a wheelchair. Disabled people can efficiently operate the wheelchair as no prior training is necessary. In addition, the smart glass is lightweight to wear. As their is less computational complexity in this system, its response time and performance are sufficient for real-time uses. This technology gives a disabled person the opportunity to feel independent to move anywhere. It is not easy to detect eye pupils in dark places, so, sufficient light is needed for detecting eye movement. In addition, the angle by which the light is falling on the camera also plays a vital role. The user should ensure the proper positioning of the glasses on the eyes. The system works perfectly in minimum vibration. Consequently, less fluctuation of eyes is desirable. Training a large data set makes this project robust [10] for different light intensities falling on the eye from the glass. Some sensors can also be used for helping disabled people. The ultrasonic sensor can be added to avoid obstacles, and an emergency switch can also be added for any emergency.

ACKNOWLEDGEMENT

This work was performed by a group of members from the Robotics and Automation Laboratory of Chittagong University of Engineering and Technology. Support was also provided by the Department of Mechatronics and Industrial Engineering. We thank Md. Anisur Rahman for help in gathering reference material and assisting in making the prototype.

DISCLOSURE STATEMENT

The authors declare no conflicts of interest.

Bibliography

[1] Rani, M.S., Chitransh, S., Tyagi, P. and Varshney, P., 2020. Eye Controlled Wheel Chair. *International Journal of Scientific Research & Engineering Trends*, 6(3), 1821-1823.

[2] Bai, D., Liu, Z., Hu, Q., Yang, J., Yang, G., Ni, C., . . . Zhou, L. (2017). Design of an eye movement-controlled wheelchair using Kalman filter algorithm. 2016.

IEEE International Conference on Information and Automation, IEEE ICIA 2016 (August), 1664-1668.

[3] Card, S. (n.d.). Visual search of computer command menus. attention and performance X: Control of language processes (December), 97-108.

[4] Dhaval Pimplaskar Atul Borkar, M. S. N. (2013). Real time eye blinking detection and tracking using Opencv. Journal of Engineering Research and Application, 3(5), 1780-1787.

[5] Disability Inclusion Overview. (n.d.). Retrieved 2021-07-14, from https://www.worldbank.org/en/topic/disability2.

[6] Dodge, R., & Cline, T. S. (1901). The angle velocity of eye movements. Psychological Review, 8(2).

[7] Gupta, R., Kori, R., Hambir, S., Upadhayay, A. and Sahu, S., 2020, April. Eye Controlled Wheelchair Using Raspberry Pi. In Proceedings of the 3rd International Conference on Advances in Science & Technology (ICAST).

[8] Hartridge, H. and Thomson, L.C., 1948. Methods of investigating eye movements. The British journal of ophthalmology, 32(9), p. 581.

[9] Jacob, R. J., & Karn, K. S. (2003). Eye Tracking in Human-Computer Interaction and Usability Research. Ready to Deliver the Promises. In The mind's eye, (pp. 573-605)

[10] Jafar, F., Fatima, S. F., Mushtaq, H. R., Khan, S., Rasheed, A., & Sadaf, M. (2019). Eye controlled wheelchair using transfer learning. RAEE 2019 - International Symposium on Recent Advances in Electrical Engineering, 4, 1-5.

[11] Judd-1905-introduction to a series.pdf. (n.d.), from https://hdl.handle.net/11858/00-001M-0000-002C-0465-A.

[12] Lambert, R. H., Monty, R. A., & Hall, R. J. (1974). High-speed data processing and unobtrusive monitoring of eye movements. Behavior Research Methods Instrumentation, 6(6), 525-530.

[13] Lontis, E. R., Bentsen, B., Gaihede, M., & Andreasen Struijk, L. N. (2014). Wheelchair control with the tip of the tongue. Biosystems and Biorobotics, 7, 521-527.

[14] Nishimori, M., Saitoh, T., & Konishi, R. (2007). Voice controlled intelligent wheelchair. Proceedings of the SICE Annual Conference, 336-340.

[15] Patel, S. N., & Prakash, V. (2015). Autonomous camera based eye controlled wheelchair system using raspberry-pi. ICIIECS 2015 - 2015 IEEE International Conference on Innovations in Information, Embedded and Communication Systems, 3-8.

[16] Swee, S. K., Kiang, K. D. T., & You, L. Z. (2016). EEG controlled wheelchair. *MATEC Web of Conferences*, 51, 1-9.

[17] Tsutomu, O. (1994). Thermal effect of visible light and infra-red radiation (i.r.-A, i.r.-B and i.r.-C) on the eye: A study of infra-red cataract based on a model. *Annals of Occupational Hygiene*, 38(4), 351-359.

Index